EDINBURGH'S TELFORD COLLEGE

0042449

KT-482-451

Scottish Examination Materials

Test Your
HIGHER
CHEMISTRY
Calculations

SECOND EDITION

007

David Calder

07

Hodder Gibson

2A Christie Street, Paisley, PA1 1NB

	16/6/05			611
		004244449		
		£ 7.99		
		540.76 NPH		
	SC	NCLF	2 WK LOAN
22/6/05	MW		OU

Although every effort has been made to ensure that website addresses are correct at time of going to press, Hodder Gibson cannot be held responsible for the content of any website mentioned in this book. It is sometimes possible to find a relocated web page by typing in the address of the home page for a website in the URL window of your browser.

Papers used in this book are natural, renewable and recyclable products. They are made from wood grown in sustainable forests. The logging and manufacturing processes conform to the environmental regulations of the country of origin.

While every effort has been made to check the instructions of practical work in this book, it is still the duty and legal obligation of schools to carry out their own risk assessments.

Orders: please contact Bookpoint Ltd, 130 Milton Park, Abingdon, Oxon OX14 4SB. Telephone: (44) 012135 827720. Fax: (44) 01235 400454. Lines are open from 9.00–6.00, Monday to Saturday, with a 24-hour message answering service. Visit our website at www.hodderheadline.co.uk. Hodder Gibson can be contacted direct on: Tel: 0141 848 1609; Fax: 0141 889 6315; email: hoddergibson@hodder.co.uk

© David Calder 2004
First Edition Published 1997
Second Edition Published 2004
This Edition Published 2004 by
Hodder Gibson, a member of the Hodder Headline Group
2a Christie Street
Paisley PA1 1NB

Impression number   10 9 8 7 6 5 4 3 2 1
Year   2010 2009 2008 2007 2006 2005 2004

All rights reserved. Apart from any use permitted under UK copyright law, no part of this publication (except those pages indicated) may be reproduced or transmitted in any form or by any means, electronic or mechanical, including photocopy, recording, or any information storage and retrieval system, without permission in writing from the publisher or under licence from the Copyright Licensing Agency Limited. Further details of such licences (for reprographic reproduction) may be obtained from the Copyright Licensing Agency Limited, of 90 Tottenham Court Road, London W1T 4LP.

Typeset by Tech-set Ltd.
Printed in Great Britain for Hodder Gibson, a division of Hodder Headline, 2A Christie Street, Paisley PA1 1NB by J.W. Arrowsmith Ltd, Bristol.

A catalogue record for this title is available from the British Library.

ISBN 0 340 883 154

Contents

The Mole: Masses and Concentrations of Solutions

This chapter is in two parts; Part 1 deals with the definition of the mole as a formula mass expressed in grams and Part 2 involves the concentrations of solutions expressed in mol l^{-1} ('moles per litre'). This is a revision of work done at Standard Grade or Intermediate 2 Chemistry.

PART 1: The Mole as a Mass

The definition of a mole as the formula mass of a substance, expressed in grams, should be familiar; this applies whether the substance is an element or a compound. Note that the accepted abbreviation for the mole is **mol**. For example:

$$\begin{aligned}
1 \text{ mole of Cu} &= 63.5 \text{ g} \\
1 \text{ mole of } H_2 &= 2 \text{ g} \\
1 \text{ mole of NaCl} &= 58.5 \text{ g} \\
1 \text{ mole of } CO_2 &= 44 \text{ g}
\end{aligned}$$

The worked examples and problems in this chapter should be easy but should not be seen as trivial. They are intended as revision of earlier work and to give practice in setting out arithmetical problems clearly.

Worked Example 1.1

Calculate the mass of 2.5 mol of copper.

> The relative atomic mass of Cu = 63.5.
> So 1 mol of Cu = 63.5 g
> So 2.5 mol of Cu = 2.5 × 63.5 g
> = 158.75 g

Note that this initial statement is written this way round, rather than '63.5 g of Cu = 1 mol' since our answer has to be a number of grams, which we want on the right hand side of the problem.

Worked Example 1.2

How many moles of sodium chloride are present in 11.7 g of the salt?

Formula of sodium chloride is NaCl.
Formula mass of NaCl = 23 + 35.5 = 58.5.
So 1 mole = 58.5 g

Since we want our answer to come out as a number of moles on the right hand side of the problem, we turn the above statement round to give:

$$58.5 \, \text{g} = 1 \, \text{mol of NaCl}$$

$$1 \, \text{g} \quad = \frac{1}{58.5} \, \text{mol of NaCl}$$

$$11.7 \, \text{g} = \frac{11.7}{58.5} \, \text{mol of NaCl}$$

$$= 0.2 \, \text{mol of NaCl}$$

PROBLEMS

These problems are of the type shown in Worked Examples 1.1 and 1.2.

1.1 What is the mass of 0.2 mol of calcium nitrate, $Ca(NO_3)_2$?

1.2 How many moles of aluminium carbonate, $Al_2(CO_3)_3$ are present in 4.68 g of the substance?

1.3 How many moles of ammonium carbonate, $(NH_4)_2CO_3$ are present in 115.2 g of the substance?

1.4 What is the mass of 0.025 mole of sucrose, $C_{12}H_{22}O_{11}$?

1.5 Washing soda, sodium carbonate–10–water, has the formula $Na_2CO_3.10H_2O$. How many moles of washing soda are present in 0.715 g of the substance?

1.6 What is the mass of 0.4 mol of ammonium phosphate, $(NH_4)_3PO_4$?

1.7 How many moles of water, H_2O, are present in 5.4 g of the substance?

1.8 What is the mass of 0.08 mol of carbon monoxide, CO?

1.9   How many moles of barium chloride, $BaCl_2$, are present in 4.166 g of the salt?

1.10   What is the mass of 1.2 mol of sodium hydroxide, NaOH?

1.11   What is the mass of 0.2 mol of copper(II) chloride, $CuCl_2$?

1.12   How many moles of nitric acid, HNO_3, are present in 94.5 g of the pure substance?

1.13   What is the mass of 0.025 mol of iron(III) oxide, Fe_2O_3?

1.14   How many moles of silver(I) nitrate, $AgNO_3$, are present in 6.796 g of the substance?

1.15   What is the mass of 3.5 mol of propene, C_3H_6?

1.16   How many moles of ammonium sulphate, $(NH_4)_2SO_4$, are present in 105.68 g of the substance?

1.17   What is the mass of 0.5 mol of copper(II) sulphate–5–water, $CuSO_4.5H_2O$?

1.18   How many moles of mercury(II) nitrate, $Hg(NO_3)_2$, are present in 16.23 g of the substance?

1.19   What is the mass of 0.3 mol of sodium carbonate Na_2CO_3?

1.20   How many moles of aluminium hydrogensulphite, $Al(HSO_3)_3$, are present in 8.109 g of the substance?

PART 2: The Mole and Concentration

The concentration of a solution in Chemistry is expressed as the number of moles of substance dissolved in each litre of the solution. In the Standard Grade or Intermediate 2 courses, the abbreviated unit was mol/litre ('moles per litre'). In the Higher course, the abbreviation used is mol l^{-1}, which has an identical meaning. Although not used in this book or in SQA examinations the term 'Molar', abbreviated M, which also has the same meaning, will often be found in other textbooks and in the labelling of containers of solutions.

In arithmetical terms we describe concentration by the equation:

$$\text{concentration} = \frac{\text{number of moles}}{\text{volume (in litres)}}$$

This is often shown in the form of a triangle, as shown below:

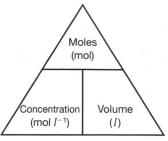

As can be seen, the three quantities are in the same positions in the triangle as they are in the above equation, that is, concentration on the left, moles above and litres on the right. This can be helpful in obtaining the other two equations connecting these quantities – they are:

$$\text{volume (in litres)} = \frac{\text{number of moles}}{\text{concentration}}$$

$$\text{number of moles} = \text{volume (in litres)} \times \text{concentration}$$

As can be seen, the positions of the three quantities are the same as in the triangle. So only the original equation or triangle needs to be remembered to be able to work out the other equations.

Worked Example 1.3

What is the concentration of a solution containing 2.5 mol of substance dissolved in 5 l of solution?

The appropriate equation is selected and the data put in:

$$\text{concentration} = \frac{\text{number of moles}}{\text{volume (in litres)}}$$

$$= \frac{2.5}{5}$$

$$= 0.5 \text{ mol } l^{-1}$$

Worked Example 1.4

How many moles of substance are present in 25 cm^3 of a 0.2 mol l^{-1} solution?

In this problem, the volume of solution has been given in cm^3 and must firstly be converted to litres.

In most laboratory Chemistry, only small volumes of solutions are used in practice. So the smaller unit of the **cubic centimetre**, abbreviated **cm^3**, is commonly used.

$$1 \text{ litre} = 1000 \text{ cm}^3$$

So
$$1 \text{ cm}^3 = \frac{1}{1000} \text{ litre} = 0.001 \text{ litre}$$

Note that the term **millilitre**, abbreviated **ml**, is often used as a unit of volume for measuring household liquid volumes, e.g. drinks, detergents etc. This unit used to have a slightly different definition from the cubic centimetre, but nowadays the units are identical. However, the millilitre is not used in the SQA exams and will not be used in this book.

Since $1 \, l = 1000 \text{ cm}^3$, 25 cm^3 is $25/1000 \, l = 0.025 \, l$. This value can now be fitted into the appropriate equation:

$$\text{No. of moles} = \text{concentration} \times \text{volume (in litres)}$$
$$= 0.2 \times 0.025$$
$$= \mathbf{0.005 \text{ mol}}$$

Worked Example 1.5

2 g of sodium hydroxide, NaOH, are dissolved in water to make 0.4 mol l^{-1} solution. What volume is the solution?

In order to calculate the volume we need to know the concentration and the number of moles, but we are only told the concentration; we must first work out the number of moles of sodium hydroxide from its mass:

Formula of sodium hydroxide: NaOH.
Formula mass: $23 + 16 + 1 = 40$.
So 1 mol of NaOH = 40 g.

Reversing this statement, to put 'mol of NaOH' on the right hand side, we have:

$$40 \text{ g} = 1 \text{ mol of NaOH}$$
$$1 \text{ g} = \tfrac{1}{40} \text{ mol of NaOH}$$
$$2 \text{ g} = \tfrac{2}{40} \text{ mol of NaOH}$$
$$= 0.05 \text{ mol of NaOH}$$

We now fit this value, and that of the concentration, into the appropriate equation:

$$\text{volume (in litres)} = \frac{\text{number of moles}}{\text{concentration}}$$
$$= \frac{0.05}{0.4}$$
$$= \mathbf{0.125 \, l \, (125 \text{ cm}^3)}$$

Worked Example 1.6

What mass of sodium carbonate, Na_2CO_3, must be dissolved to make 0.25 l of a 0.2 mol l^{-1} solution?

We have the information to calculate the number of moles of sodium carbonate required. This can be fitted into the appropriate equation directly:

$$\text{No. of moles} = \text{concentration} \times \text{volume (in litres)}$$
$$= 0.2 \times 0.25$$
$$= 0.05 \text{ mol}$$

However, the question asks for the **mass** of sodium carbonate which this represents.

Formula of sodium carbonate: Na_2CO_3,
Formula mass: $(2 \times 23) + 12 + (3 \times 16) = 106$
1 mol of sodium carbonate $= 106$ g
0.05 mol of sodium carbonate $= 0.05 \times 106$ g
  $= 5.3$ g

PROBLEMS

- Problems 1.21 to 1.25 are of the type shown in Worked Examples 1.3 and 1.4.
- Problems 1.26 to 1.40 are of the type shown in Worked Examples 1.5 and 1.6.

1.21 0.24 mol of salt is dissolved to make 1.2 l of solution. What is the concentration of the solution?

1.22 200 cm^3 of a salt solution has a concentration of 1.5 mol l^{-1}. How many moles of salt are dissolved in it?

1.23 0.005 mol of a substance is dissolved in 25 cm^3 of solution. What is the concentration of the solution?

1.24 What is the volume (in cm^3) of a 1.2 mol l^{-1} solution which contains 0.048 mol of dissolved substance?

1.25 A flask contains 300 cm^3 of a 0.5 mol l^{-1} acid solution. How many moles of pure acid must have been dissolved?

1.26 2.943 g of pure sulphuric acid, H_2SO_4, is dissolved in water to make 150 cm^3 of solution. What is the concentration of the acid solution now?

1.27 A 0.2 mol l^{-1} solution of sodium carbonate, Na_2CO_3, is made by dissolving 5.3 g of the solid in water and making it up to the mark in a standard flask. What volume must the flask be?

1.28 0.8 l of a 0.5 mol l^{-1} solution of ammonium nitrate, NH_4NO_3, has to be made up; what mass of solid ammonium nitrate would be required?

1.29 71.05 g of sodium sulphate, Na_2SO_4, is dissolved to make a 2 l standard solution. What concentration is the solution?

1.30 25 cm^3 of a 0.4 mol l^{-1} solution of ammonium sulphate, $(NH_4)_2SO_4$, is made up. What mass of solid must have been dissolved?

1.31 A 0.2 mol l^{-1} solution of potassium nitrate, KNO_3, is made by dissolving 30.33 g of the solid in a standard flask. What volume is the standard flask?

1.32 0.4 g of sodium hydroxide, NaOH, is dissolved in water to make a 0.25 mol l^{-1} solution. What volume is the solution?

1.33 2.675 g of ammonium chloride, NH_4Cl, is dissolved in water to make a 0.1 mol l^{-1} solution. What volume of solution is made?

1.34 What mass of anhydrous copper(II) sulphate, $CuSO_4$, would be required to make 100 cm^3 of a 0.5 mol l^{-1} solution?

1.35 What would be the concentration of a 200 cm^3 solution of silver(I) nitrate, $AgNO_3$, containing 1.699 g of dissolved solid?

1.36 1.92 g of ammonium carbonate, $(NH_4)_2CO_3$, is dissolved to make 400 cm^3 of solution. What is the concentration of the solution?

1.37 2.764 g of potassium carbonate, K_2CO_3, is dissolved in 200 cm^3 of solution. What is the concentration of the solution?

1.38 What mass of pure ethanoic acid, CH_3COOH, would be required to make 40 cm^3 of a 0.4 mol l^{-1} solution?

1.39 14.3 g of sodium carbonate–10–water, $Na_2CO_3.10H_2O$, is dissolved in water and made up to 250 cm^3 in a standard flask. What is the concentration of the sodium carbonate solution?

1.40 Oxalic acid has the formula $(COOH)_2$. If 0.0225 g of the pure acid was dissolved in water to make a 0.01 mol l^{-1} solution, what would the volume of the solution be?

Calculations from Equations – 1: Involving Masses Only

Calculations in which we are told the quantity of one of the chemical species in a reaction and asked to work out the quantity of another species can be solved by the use of the methods described below. The work of this chapter contains some work which is mainly a revision of Standard Grade or Intermediate 2 Chemistry; however, such problems should not be regarded as being trivial as many of them are of a type which could appear in the Higher examination. The method outlined may seem unnecessarily long, but only because **all** working and reasoning is shown. Although there **are** 'short cuts', these do not actually involve a different **method**; they merely save a very small amount of time by omitting to show all the working. Such short cuts are not recommended until the method has been completely mastered.

Worked Example 2.1

What mass of copper would be obtained by heating 3.975 g of copper(II) oxide with an excess of carbon?

Step 1: Balanced Equation

$$2CuO + C \longrightarrow 2Cu + CO_2$$

Step 2: Mole Statement
The purpose of this is to connect the number of moles of what you are told about (the 'known' species) with the number of moles of what you are asked about (the 'unknown' species). This is obtained directly from the balanced equation.

2 mol of CuO   produce   2 mol of Cu
So   1 mol of CuO   produces   1 mol of Cu

Step 3: Calculation of 'known moles'
In this step we calculate the number of moles of the 'known' substance. In this example, we are told the **mass of copper(II) oxide**. So we calculate what 1 mol of copper(II) oxide weighs and work out from that how many moles 3.975 g represents.

$$1 \text{ mol of CuO} = 63.5 + 16 = 79.5$$

$$79.5\,\text{g} \quad = 1 \text{ mol of CuO}$$

$$1\,\text{g} \quad = \frac{1}{79.5} \text{ mol of CuO}$$

$$3.975\,\text{g} = \frac{3.975}{79.5} \text{ mol of CuO}$$

$$= 0.05 \text{ mol of CuO}$$

Step 4: Calculation of 'unknown moles'

In this step we refer back to the mole statement (Step 2) and use the number of 'known moles' that we have just calculated to obtain the number of 'unknown moles':

1	mol of CuO	produces	1	mol of Cu
0.05	mol of CuO	produces	**0.05**	**mol of Cu**

Step 5: Finishing Off

In this last step, we convert the number of 'unknown moles' that we have just worked out into the quantity that the question is asking about. In this example, we are asked to obtain the **mass of copper**.

$$1 \quad \text{mol of Cu} = 63.5 \qquad \text{g}$$
$$0.05 \text{ mol of Cu} = \quad 0.05 \times 63.5\,\text{g}$$
$$= \quad 3.175 \qquad \text{g}$$

PROBLEMS

These problems are simple ones of the type illustrated by Worked Example 2.1.

2.1
$$Mg + H_2SO_4 \longrightarrow MgSO_4 + H_2$$
What mass of hydrogen gas would be evolved if 6.075 g of magnesium reacted completely with dilute sulphuric acid?

2.2
$$CuO + H_2 \longrightarrow Cu + H_2O$$
What mass of copper metal would be obtained by the complete reduction of 15.9 g of copper(II) oxide by hydrogen gas?

2.3
$$2PbO + C \longrightarrow 2Pb + CO_2$$
What mass of carbon would be needed to reduce 22.32 g of lead(II) oxide completely to lead metal and carbon dioxide?

2.4
$$2CO + O_2 \longrightarrow 2CO_2$$

What mass of carbon monoxide would require to be completely burned in oxygen to form 5.5 g of carbon dioxide?

2.5
$$2Na + 2H_2O \longrightarrow 2NaOH + H_2$$

0.5 g of hydrogen gas is given off when sodium is reacted completely with water. What was the mass of sodium which reacted?

2.6
$$Zn + 2HCl \longrightarrow ZnCl_2 + H_2$$

13.08 g of zinc is reacted with excess hydrochloric acid. What mass of hydrogen would be produced?

2.7
$$Mg + 2CH_3COOH \longrightarrow Mg(CH_3COO)_2 + H_2$$

What mass of magnesium will completely react with 36 g of pure ethanoic acid, CH_3COOH?

2.8
$$Na_2CO_3 + H_2SO_4 \longrightarrow Na_2SO_4 + CO_2 + H_2O$$

5.3 g of sodium carbonate is reacted with an excess of sulphuric acid. What mass of carbon dioxide would be evolved?

2.9
$$C_6H_{12}O_6 + 6O_2 \longrightarrow 6CO_2 + 6H_2O$$

45 g of glucose is burned completely in an excess of oxygen. What mass of carbon dioxide would be produced?

2.10
$$2Ag_2CO_3 \longrightarrow 4Ag + 2CO_2 + O_2$$

What mass of oxygen would be produced by the complete thermal (by heat) decomposition of 27.58 g of silver(I) carbonate?

Worked Example 2.2

Hydrazine, N_2H_4, is a rocket fuel which reacts with oxygen producing gaseous products of nitrogen and water vapour under very high pressures. If 6.4×10^4 kg of hydrazine is burned completely, what mass of water vapour will be produced?

The only difference between this question and the previous type is that the quantity referred to is expressed as a large number of kilograms, rather than a small number of grams. This refers to a process being carried out on an industrial scale instead of using laboratory size quantities. Another unit of mass which can be involved in industrial quantities is the **tonne**, which is equal to 1000 kg. Note that this metric unit is not the same as the ton, used in the UK (but less often nowadays).

The easiest way to approach this is to work out the answer as if the question had referred to 6.4 **grams** of hydrazine; the conversion to **kilograms** can wait until the final step of the problem, as shown below.

Step 1: Balanced Equation

$$N_2H_4 + O_2 \longrightarrow N_2 + 2H_2O$$

Step 2: Mole Statement

1 mol of N_2H_4 reacts to form 2 mol of H_2O

Step 3: 'Known' Moles

'Known' substance is N_2H_4

$$32\text{ g of } N_2H_4 \;=\; 1 \quad \text{mol}$$

$$1\text{ g of } N_2H_4 \;=\; \frac{1}{32} \text{ mol}$$

$$6.4\text{ g of } N_2H_4 \;=\; \frac{6.4}{32} \text{ mol}$$

$$=\; 0.2 \;\; \text{mol of } N_2H_4$$

Step 4 'Unknown' Moles

'Unknown' is H_2O

From the Mole Statement:

$$1 \quad \text{mol of } N_2H_4 \text{ reacts to form 2 mol of } H_2O$$

So $\qquad$ 0.2 mol of N_2H_4 reacts to form **0.4 mol of H_2O**

Step 5: Finishing Off

$$1 \quad \text{mol of } H_2O = 18\text{ g}$$

So $\qquad$ 0.4 mol of $H_2O = 0.4 \times 18$ g

$$= 7.2\text{ g}$$

This would be the end of the question if it has asked us about the reaction of 6.4 g of N_2H_4; however, the question referred to 6.4×10^4 kg of N_2H_4. A simple bit of direct proportion gives us the answer:

6.4 g $\qquad$ of N_2H_4 reacts to form 7.2 g $\qquad$ of H_2O

6.4×10^4 kg of N_2H_4 reacts to form 7.2×10^4 kg of H_2O

PROBLEMS

These problems of the type illustrated by Worked Example 2.2 involve masses in 'industrial' quantities, often using 'Standard Form' e.g. 3×10^3 kg, to express masses.

2.11
$$2SO_2 + O_2 \longrightarrow 2SO_3$$
Assuming 100% conversion of reactants to products, what mass of sulphur dioxide would be required to produce 1602 kg of sulphur trioxide by the Contact Process described by the above equation?

2.12
$$N_2 + 3H_2 \longrightarrow 2NH_3$$
175 kg of nitrogen is completely converted to ammonia in the Haber Process represented by the above equation. What mass of hydrogen must have reacted?

2.13
$$Fe_2O_3 + 3CO \longrightarrow 2Fe + 3CO_2$$
63.84 tonne of iron(III) oxide is completely reduced by carbon monoxide to form pure iron. What mass of iron would be obtained?

2.14
$$2SO_2 + O_2 \longrightarrow 2SO_3$$
7692 kg of sulphur dioxide is completely converted to sulphur trioxide by the above process. What mass of sulphur trioxide would be obtained?

2.15
$$C_2H_4 + H_2O \longrightarrow C_2H_5OH$$
Ethanol C_2H_5OH, can be produced industrially by the catalytic reaction of ethene with steam. What mass of ethene would be needed to produce 1104 kg of ethanol?

2.16
$$3NO_2 + H_2O \longrightarrow 2HNO_3 + NO$$
The final stage in the industrial production of nitric acid, HNO_3, involves the above reaction. What mass of nitrogen dioxide must react to produce 2.52×10^3 kg of nitric acid?

2.17
$$C_2H_2 + H_2O \longrightarrow CH_3CHO$$
The above equation shows the catalytic hydration of ethyne, C_2H_2, to ethanal, CH_3CHO. If 2.08×10^4 kg of ethyne is completely reacted, what mass of ethanal would be produced?

2.18
$$TiCl_4 + 4Na \longrightarrow Ti + 4NaCl$$
Titanium metal can be extracted from titanium chloride by displacement by sodium. What mass of sodium would be needed to react completely with 7.596 tonne of titanium chloride?

2.19
$$N_2H_4 + 2F_2 \longrightarrow N_2 + 4HF$$

What mass of fluorine would be required to react completely with 8×10^4 kg of hydrazine, N_2H_4?

2.20
$$CH_3CHO + 3Cl_2 \longrightarrow CCl_3CHO + 3HCl$$

What mass of ethanal, CH_3CHO, would react with an excess of chlorine to produce 2.95×10^4 kg of trichloroethanal, CCl_3CHO?

3

Calculations from Equations – 2: Involving Concentrations of Solutions

This chapter involves calculations from equations where the quantity of one or both of the substances referred to is expressed in terms of the concentration and volume of a solution. The method involved is exactly the same as that used in Chapter 2 except that some quantities are expressed in terms of the volume and concentration of solutions. It is recommended that Chapters 1 and 2 are revised before starting this chapter.

Worked Example 3.1

What volume of $0.5 \, \text{mol} \, l^{-1}$ sodium hydroxide solution will exactly neutralise $40 \, \text{cm}^3$ of $0.2 \, \text{mol} \, l^{-1}$ sulphuric acid?

Step 1: Balanced Equation

$$2NaOH + H_2SO_4 \longrightarrow Na_2SO_4 + 2H_2O$$

Step 2: Mole Statement
2 mol of NaOH neutralise 1 mol of H_2SO_4

Step 3: Calculation of 'Known' Moles
In this case we are told the **volume** and **concentration** of the sulphuric acid. This enables us to calculate the number of moles.

$$\text{number of moles} = \text{volume (in litres)} \times \text{concentration}$$
$$= 0.04 \times 0.2$$
$$= 0.008 \, \text{mol (of } H_2SO_4)$$

Step 4: 'Unknown' Moles
2 mol of NaOH neutralise 1 mol of H_2SO_4 (from Step 2)
Rewriting to put our 'unknown', NaOH, on the right:

1 mol of H_2SO_4 is neutralised by 2 mol of NaOH
0.008 mol of H_2SO_4 is neutralised by 2×0.008 mol of NaOH
$$= 0.016 \, \text{mol of NaOH}$$

Step 5: Finishing Off

The problem asks for the **volume** of NaOH solution. We are told in the problem that its concentration is 0.5 mol l^{-1} and we have just calculated that we need 0.016 mol of it. So we select the correct equation:

$$\text{volume (in litres)} = \frac{\text{number of moles}}{\text{concentration}}$$

$$= \frac{0.016}{0.5}$$

$$= 0.032\, l\ (32\text{ cm}^3)$$

Worked Example 3.2

A 40 cm³ sample of sodium chloride solution is reacted with an excess of silver(I) nitrate solution, $AgNO_3$. The precipitate of silver(I) chloride formed is filtered, dried and weighed; its mass is found to be 7.17 g. Calculate the concentration of the sodium chloride solution, in mol l^{-1}.

Step 1: Balanced Equation

$$AgNO_3 + NaCl \longrightarrow AgCl + NaNO_3$$

Step 2: Mole Statement

1 mol of NaCl will produce 1 mol of AgCl precipitate

Note: The $AgNO_3$ is not required in the Mole Statement, since we are neither asked about nor told about its quantity. All we need to know is what we are told; namely that it is **in excess**; that is there is enough of it to react with **all** the available NaCl.

Step 3: Calculation of 'Known' Moles

In this case we are told the **mass** of the silver(I) chloride precipitate

$$1 \text{ mol of AgCl} = 107.9 + 35.5 = 143.4\,\text{g}$$

$$143.4\,\text{g} = 1 \quad \text{mol of AgCl}$$

$$1\,\text{g} = \frac{1}{143.4} \text{ mol of AgCl}$$

$$7.17\,\text{g} = \frac{7.17}{143.4} \text{ mol of AgCl}$$

$$= \textbf{0.05 mol of AgCl}$$

WILFORD COLLEGE LIBRARY

Step 4: 'Unknown' Moles
From Step 2, we know that:

1 mol of NaCl will produce 1 mol of AgCl precipitate.

So,

0.05 mol of NaCl will produce 0.05 mol of AgCl precipitate

That is, there must have been **0.05 mol** of NaCl in the original solution

Step 5: Finishing Off
The problem asks for the **concentration** of the NaCl solution, and tells us that its volume is $40\,cm^3$ (0.04 litres). The correct equation is selected and the data inserted:

$$\text{concentration} = \frac{\text{number of moles}}{\text{volume (in litres)}}$$

$$= \frac{0.05}{0.04}$$

$$= 1.25 \text{ mol } l^{-1}$$

PROBLEMS

These problems are of the type illustrated in Worked Examples 3.1 and 3.2 involving the concentrations and volumes of solutions. Problems 3.11–3.15 involve reagents which may be unfamiliar but where enough information is given to write a mole statement and, therefore, solve the problems using the same method as before.

3.1
$$2KOH + H_2SO_4 \longrightarrow K_2SO_4 + 2H_2O$$
$50\,cm^3$ of $0.4\,mol\,l^{-1}$ potassium hydroxide solution exactly neutralises $20\,cm^3$ of sulphuric acid. What concentration was the acid?

3.2
$$2NaOH + H_2SO_4 \longrightarrow Na_2SO_4 + 2H_2O$$
$20\,cm^3$ of $0.2\,mol\,l^{-1}$ sulphuric acid exactly neutralises a quantity of $0.5\,mol\,l^{-1}$ sodium hydroxide solution. What volume of the alkali must have been reacted?

3.3
$$2KOH + CO_2 \longrightarrow K_2CO_3 + H_2O$$
$200\,cm^3$ of $0.5\,mol\,l^{-1}$ potassium hydroxide solution is reacted with an excess of carbon dioxide. What mass of potassium carbonate would be obtained upon complete evaporation of the water from the resulting solution?

3.4
$$H_3PO_4 + 3NaOH \longrightarrow Na_3PO_4 + 3H_2O$$
$1.96\,g$ of pure phosphoric acid, H_3PO_4, is dissolved in water and the solution is exactly neutralised by a quantity of $0.2\,mol\,l^{-1}$ sodium hydroxide solution. What volume of the alkali solution must have been required?

3.5
$$2NH_3 + H_2SO_4 \longrightarrow (NH_4)_2SO_4$$
6.8 g of ammonia gas is required to neutralise exactly 250 cm³ of sulphuric acid. What concentration must the acid have been?

3.6
$$C_6H_4(COOH)_2 + 2NaOH \longrightarrow C_6H_4(COONa)_2 + 2H_2O$$
A solution of phthalic acid, $C_6H_4(COOH)_2$, is exactly neutralised by 25 cm³ of a 2 mol l^{-1} solution of sodium hydroxide. What mass of phthalic acid must have been dissolved in the original solution?

3.7
$$CH_3COOH + NaOH \longrightarrow CH_3COONa + H_2O$$
A 20 cm³ sample of vinegar (dilute ethanoic acid, CH_3COOH) is titrated with standard 0.1 mol l^{-1} sodium hydroxide solution. Exactly 36.4 cm³ of the alkali is found to neutralise the acid sample. What is the concentration, in mol l^{-1}, of the acid in the vinegar sample?

3.8
$$H_2SO_4 + BaCl_2 \longrightarrow BaSO_4 + 2HCl$$
12.5 cm³ of 0.4 mol l^{-1} solution of sulphuric acid is poured into a solution containing an excess of barium chloride. A precipitate of barium sulphate is formed which is filtered, washed, dried and weighed. What mass of precipitate should theoretically be obtained?

3.9
$$2KOH + (COOH)_2 \longrightarrow (COOK)_2 + 2H_2O$$
A solution of oxalic acid, $(COOH)_2$, is standardised by taking a 25 cm³ sample of it and titrating it against 0.2 mol l^{-1} potassium hydroxide solution. Neutralisation is obtained by the addition of 23.5 cm³ of the alkali. Calculate the concentration, in mol l^{-1}, of the acid.

3.10
$$Pb(NO_3)_2 + H_2SO_4 \longrightarrow PbSO_4 + 2HNO_3$$
20 cm³ of dilute sulphuric acid is reacted with an excess of lead(II) nitrate, causing 1.2132 g of lead(II) sulphate to be precipitated. Calculate the concentration, in mol l^{-1}, of the acid.

3.11 A compound known as ethylenediaminetetraacetic acid (EDTA) is useful for measuring the quantities of certain metal ions in solution. For example, Ca^{2+} ions and EDTA react in a 1 mol : 1 mol ratio. It is found that 14.6 cm³ of 0.1 mol l^{-1} EDTA solution reacts exactly with a 25 cm³ sample of a solution containing Ca^{2+} ions. Calculate the concentration, in mol l^{-1}, of the calcium ion solution.

3.12 Potassium permanganate, in the presence of acid, reacts with sulphite ions in the ratio 1 mol permanganate : 2.5 mol sulphite. When all the permanganate has reacted, its purple colour disappears and the solution becomes clear. Thus, standard acidified potassium permanganate solution can be easily used to measure the concentration of sulphite ions in a solution.

$17 \, cm^3$ of standard $0.1 \, mol \, l^{-1}$ acidified potassium permanganate solution reacts exactly with $21.25 \, cm^3$ of a solution containing sulphite ions. Calculate the concentration, in $mol \, l^{-1}$, of the sulphite ion solution.

3.13 Potassium dichromate, in the presence of acid, reacts with Fe^{2+} ions in the ratio 1 mol potassium dichromate : 6 mol Fe^{2+}. $25 \, cm^3$ of a solution containing Fe^{2+} ions was titrated with $0.1 \, mol \, l^{-1}$ acidified potassium dichromate solution. It was found that $15 \, cm^3$ of the dichromate solution exactly reacted with all the Fe^{2+} ions. Calculate the concentration, in $mol \, l^{-1}$, of the Fe^{2+} solution.

3.14 EDTA solution reacts in a 1 mol : 1 mol ratio with Ni^{2+} ions under certain conditions. A $20 \, cm^3$ sample of nickel(II) solution was titrated with $0.05 \, mol \, l^{-1}$ EDTA solution; $48 \, cm^3$ of the EDTA was required to react exactly with the Ni^{2+} ions. What concentration, in $mol \, l^{-1}$, is the nickel(II) solution?

3.15 Standard acidified potassium permanganate can be used to determine the concentration of hydrogen peroxide solution; the solutions react in the ratio 1 mol of potassium permanganate : 2.5 mol of hydrogen peroxide. In an analysis it is found that $16.8 \, cm^3$ of standard $0.025 \, mol \, l^{-1}$ potassium permanganate solution reacts exactly with a $50 \, cm^3$ sample of hydrogen peroxide solution. What is the concentration, in $mol \, l^{-1}$, of the hydrogen peroxide solution?

Worked Example 3.3

A sample of sulphuric acid is to analysed by titration with sodium carbonate solution. A $25 \, cm^3$ sample of it is pipetted into a $250 \, cm^3$ standard flask and the solution is made up to the mark. A $25 \, cm^3$ sample of this diluted solution is titrated with standard $0.2 \, mol \, l^{-1}$ sodium carbonate solution; it is found that $15 \, cm^3$ of the sodium carbonate solution is required to neutralise the acid sample. What concentration was the *original (undiluted)* acid.

In problems which involve samples being diluted it is difficult to generalise about the best method. However, usually, dilutions are made in multiples of 10; when this is the case the problem can usually be solved without too much difficulty. In this problem, a $25 \, cm^3$ sample is diluted to $250 \, cm^3$; in other words *it has been diluted to 1/10th of its original concentration*. With that in mind, we can firstly work out the concentration of the *diluted* acid, and simply multiply this value by 10 to get the original concentration.

Step 1: Balanced Equation

$$Na_2CO_3 + H_2SO_4 \longrightarrow Na_2SO_4 + CO_2 + H_2O$$

Step 2: Mole Statement

1 mol of Na_2CO_3 neutralises 1 mol of H_2SO_4

Step 3: 'Known' Moles
In this case we are told the **volume** and **concentration** of the sodium carbonate solution. This enables us to calculate the number of moles.

$$\text{number of moles} = \text{number of litres} \times \text{concentration}$$
$$= 0.015 \times 0.02$$
$$= 0.003 \text{ mol (of } Na_2CO_3)$$

Step 4: 'Unknown' Moles
$$1 \quad \text{mol of } Na_2CO_3 \text{ is neutralised by } 1 \quad \text{mol of } H_2SO_4$$
$$0.003 \quad \text{mol of } Na_2CO_3 \text{ is neutralised by } \textbf{0.003} \text{ mol of } H_2SO_4$$

Step 5: Finishing Off
We have to find the concentration of the diluted H_2SO_4 first, so we select the correct equation:

$$\text{concentration} = \frac{\text{number of moles}}{\text{volume (in litres)}}$$
$$= \frac{0.003}{0.025}$$
$$= 0.12 \text{ mol } l^{-1}$$

This is the concentration of the *diluted* acid. Since this acid solution was made by diluting the original acid to 1/10th of its concentration, the original acid has 10 times the concentration of the diluted sample. The original acid therefore has a concentration of 10×0.12 mol l^{-1}, that is **1.2 mol l^{-1}**.

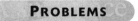 **PROBLEMS**

These problems involve the dilution of solutions as in Worked Example 3.3.

3.16
$$2NaOH + H_2SO_4 \longrightarrow Na_2SO_4 + 2H_2O$$

A 25 cm³ sample of sulphuric acid of unknown concentration is pipetted into a 250 cm³ standard flask and made up to the mark with water. A 25 cm³ sample of the new, *diluted*, solution is titrated with 0.2 mol l^{-1} sodium hydroxide solution. 50 cm³ of the alkali exactly neutralises this sample. Calculate:

(*a*)   the concentration of the *diluted* acid, in mol l^{-1};

(*b*)   how many times more concentrated the original sample is in comparison to the diluted sample;

(*c*)   the concentration of the *original acid* (using your answers to *(a)* and *(b)* above).

3.17 $$CH_3COOH + KOH \longrightarrow CH_3COOK + H_2O$$

30 g of pure ethanoic acid, CH_3COOH, is diluted with water to make up a 250 cm^3 standard solution. A 10 cm^3 sample of this diluted acid is withdrawn and titrated with 0.2 mol l^{-1} potassium hydroxide solution. Calculate:

(a)   the concentration, in mol l^{-1}, of the *diluted* acid;

(b)   the volume of alkali required to neutralise the 10 cm^3 sample.

3.18 $$H_2SO_4 + Na_2CO_3 \longrightarrow Na_2SO_4 + CO_2 + H_2O$$

19.62 g of pure sulphuric acid was weighed and made up to 1 l with water in a standard flask. A 40 cm^3 sample of this diluted solution was found to be neutralised exactly by 25 cm^3 of sodium carbonate solution. Calculate:

(a)   the concentration, in mol l^{-1}, of the diluted acid;

(b)   the concentration of the sodium carbonate solution.

3.19 $$2NaOH + H_2SO_4 \longrightarrow Na_2SO_4 + 2H_2O$$

Car batteries contain sulphuric acid. A 10 cm^3 sample of battery acid is pipetted into a 1 l standard flask and made up to the mark with water. A 50 cm^3 sample of the new diluted acid is titrated against standard 0.2 mol l^{-1} sodium hydroxide solution. It is found that 12.5 cm^3 of the alkali exactly neutralises the diluted acid sample. Calculate:

(a)   how *many times* more concentrated is the battery acid compared to the diluted acid;

(b)   the concentration of the diluted acid;

(c)   use your answers to *(a)* and *(b)* to calculate the concentration of the battery acid, in mol l^{-1}.

3.20 $$2HCl + Na_2CO_3 \longrightarrow 2NaCl + CO_2 + H_2O$$

A standard 1 mol l^{-1} solution of sodium carbonate is made up. A 20 cm^3 sample is then withdrawn from it and pipetted into a 500 cm^3 standard flask which is made up to the mark with water. A 20 cm^3 sample of this *diluted solution* is titrated against hydrochloric acid of unknown concentration. 40 cm^3 of acid exactly neutralises the sample. Calculate:

(a)   How many moles of sodium carbonate are present in the first 20 cm^3 sample.

(b)   What is the concentration of the diluted sodium carbonate solution?

(c)   What is the concentration of the hydrochloric acid?

Rate of Reaction Graphs

It has always been of interest to chemists to be able to measure the rate (speed) of chemical reactions. In this chapter we consider two ways of interpreting graphical information about reaction rates. In Part 1, information about the mass, concentration or volume of a substance during a reaction is plotted against time. The resulting curve shows the progress of the reaction; from the gradient (slope) of the graph we can obtain a measure of the average rate of reaction over any time interval. In Part 2, 'clock reactions' are introduced; when two reactant solutions are mixed, nothing appears to happen for some time and then, suddenly, a colour change takes place. Such reactions are very useful in studying the rates of reactions, but a different method of analysis is used. In these reactions the reciprocal of the time taken for the colour change (1/t) is used as a measure of the rate of the reaction.

PART 1

The progress of a chemical reaction can be followed by measuring a change in the quantity or concentration of a substance at various times as the reaction takes place. For example, in a reaction where a gas is given off, we could record the loss of mass due to escaping gas using a balance, or measure the volume of gas produced with a gas syringe or other device. The sort of apparatus used is shown below.

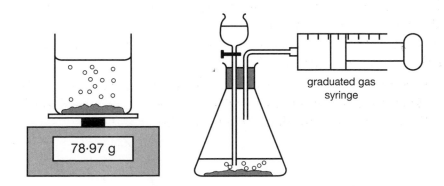

graduated gas
syringe

78·97 g

The graphs which would be obtained from these two experiments would be similar to those below.

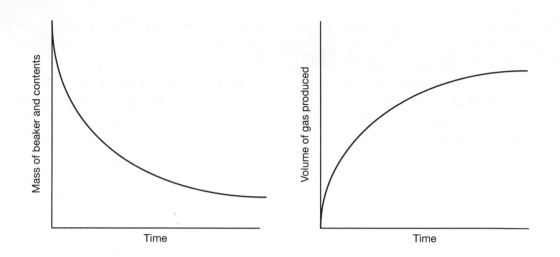

The former graph shows the mass of the beaker and contents dropping as time goes on; the latter graph shows the volume of gas being collected increasing with time. Both graphs start off with a steep gradient (slope), showing a fast reaction at first. As time goes on, the gradients get less steep as the reaction slows down (because the reactants are being used up). Eventually, the gradients reach a point where the graphs become level; that is where the gradient is zero. At this point, no more gas is being produced; in the examples considered, we can say that the reaction has stopped.

We could similarly measure the change in concentration of a reactant or product during a reaction. This might be done by using an electronic meter, such as a pH meter which gives a measure of the concentration of acid or alkali. Another method is to extract small samples of reaction mixture at different times and analyse them, for example by titration. The precise method for following any particular reaction depends on the nature of the reagents involved. The graphs obtained would look like those below.

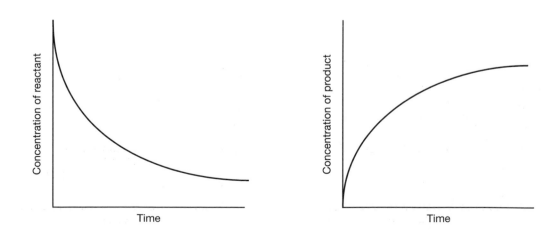

As in the previous examples, both graphs show a fast initial rate of reaction, represented by the steep slope of the graphs at the start. As time goes on, the graphs level off as the reaction slows down. Note that in the first of the two graphs, the concentration of reactant does not necessarily drop to zero, although it will eventually reach a constant value. It might be the case that the reactant being measured was in **excess**; that is, there was more of it than would completely react. Alternatively, while reactants are changing into products, **a back reaction**, in which products change into reactants, may be occurring. When these two reactions take place at the same rate, an **equilibrium** is established. For this reason, we cannot necessarily say that the reaction has 'stopped'; we *can* conclude that the **overall** rate of reactants being used up (or products being produced) has reached zero.

We can obtain a numerical value for the rate of reaction over any period of time from a graph such as those shown, using the formula below:

$$\text{rate} = \frac{\text{change in mass } \textbf{or} \text{ volume of gas } \textbf{or} \text{ concentration}}{\text{time interval over which change took place}}$$

Before seeing how this works in practice, it is important to consider the **units** in which rate of reaction may be measured. The unit of rate is simply the unit in which the quantity of substance was measured, **divided by** the unit of time used. Using the currently accepted notation, 'divided by seconds' is represented as s^{-1}, and 'divided by minutes' is min^{-1}.

Consider the graphs below which show the progress of different reactions against time. The unit for the rate of reaction is noted within the graph in each case.

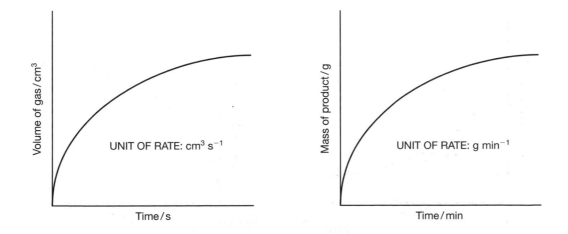

UNIT OF RATE: $cm^3\ s^{-1}$

UNIT OF RATE: $g\ min^{-1}$

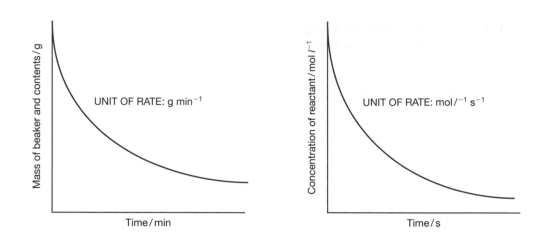

The calculation of rate of reaction from graphs such as those above is now considered in the following Worked Examples.

Worked Example 4.1

The graph below shows the mass of carbon dioxide given off in a chemical reaction, against time.

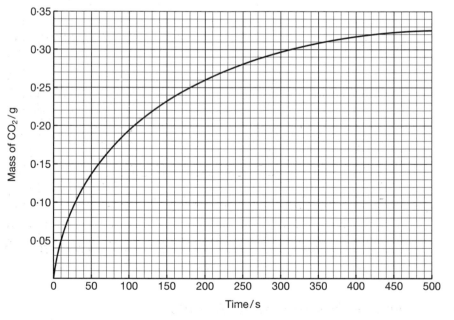

Calculate
(a) the average rate of reaction over the first 300 s;
(b) the average rate of reaction between 100 s and 400 s.

(a) The rate of reaction is defined in this case by:

$$\text{rate} = \frac{\text{change in mass}}{\text{time interval}}$$

At time $=$ 0 s, mass $= 0$ g.
At time $= 300$ s, mass $= 0.30$ g.

So: $\quad$ rate $= \dfrac{0.30 - 0}{300 - 0} = \dfrac{0.30}{300}$

$\qquad\qquad = 0.001 \text{ g s}^{-1}\ (1 \times 10^{-3} \text{ g s}^{-1})$

(*b*) $\quad$ At time $= 100$ s, mass $= 0.20$ g.
At time $= 400$ s, mass $= 0.32$ g.

So: $\quad$ rate $= \dfrac{0.32 - 0.20}{400 - 100} = \dfrac{0.112}{300}$

$\qquad\qquad = 0.0004 \text{ g s}^{-1}\ (4 \times 10^{-4} \text{ g s}^{-1})$

Worked Example 4.2

The graph below shows the change in concentration of acid against time as a chemical reaction progresses.

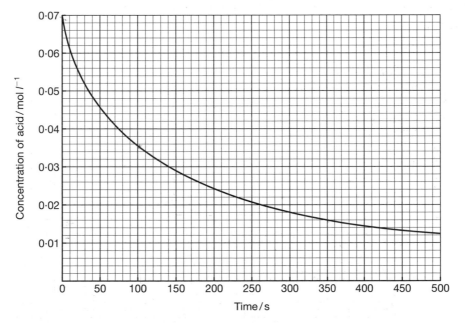

Calculate

(*a*) $\quad$ the average rate of reaction over the first 100 s;

(*b*) $\quad$ the average rate of reaction between 150 s and 350 s.

(*a*) $\quad$ The rate of reaction is defined in this case by:

$$\text{rate} = \frac{\text{change in concentration}}{\text{time interval}}$$

At time $= 0$ s, $\quad$ concentration $= 0.07$ mol l^{-1}.

At time = 100 s, concentration = 0.036 mol l^{-1}.

So: rate = $\dfrac{0.07 - 0.036}{100 - 0} = \dfrac{0.034}{100}$

$\qquad\qquad$ = 0.00034 mol $l^{-1}\,s^{-1}$ (3.4 × 10^{-4} mol $l^{-1}\,s^{-1}$)

NOTE: The value for 'change in concentration' was obtained by taking the smaller concentration from the larger to give a **positive** value. Although in mathematics, the graph shown has a negative **gradient**, a negative **rate of reaction** has no meaning. Always subtract smaller values from larger ones to obtain a positive value for the rate, even when the graph has a downward (negative) gradient.

(b)   The answer to the second part of the question is similarly calculated.

At time = 150 s, concentration = 0.03 mol l^{-1}.
At time = 350 s, concentration = 0.018 mol l^{-1}.

So: rate = $\dfrac{0.03 - 0.018}{350} - 150 = \dfrac{0.012}{200}$

$\qquad\qquad$ = 0.00006 mol $l^{-1}\,s^{-1}$ (6 × 10^{-5} mol $l^{-1}\,s^{-1}$)

PROBLEMS

Problems 4.1–4.10 are of the type shown in Worked Examples 4.1 and 4.2.

4.1   The graph below shows the mass of carbon dioxide given off against time in a chemical reaction.

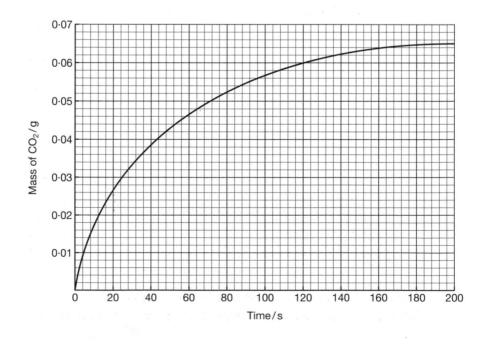

(*a*) Calculate the average rate of reaction during the first 120 s.

(*b*) Calculate the average rate of reaction between 80 s and 160 s.

4.2 The graph below shows the change in concentration of a reactant against time during a chemical reaction.

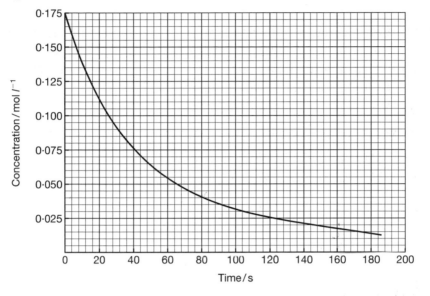

(*a*) Calculate the average rate of reaction in the first 20 s of the reaction.

(*b*) Calculate the average rate of reaction between 80 s and 160 s.

4.3 The graph below shows how the volume of hydrogen produced in a chemical reaction varies with time.

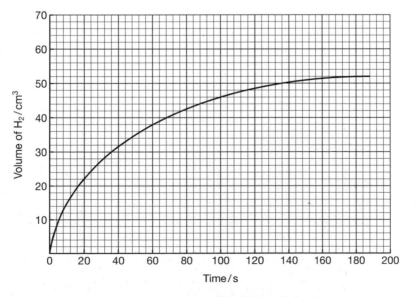

(*a*) Calculate the average rate of reaction in the first 20 s.

(*b*) Calculate the average rate of reaction between 60 s and 100 s.

4.4 The graph below shows the change in concentration of a reactant, with time, during the progress of a chemical reaction.

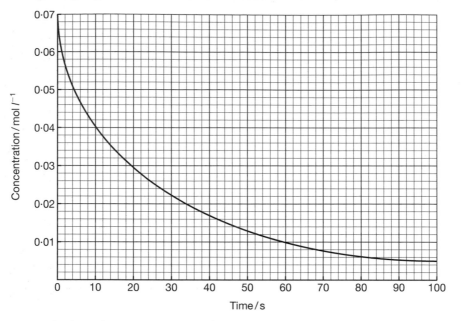

(*a*) Calculate the average rate of reaction over the first 20 s.
(*b*) Calculate the average rate of reaction between 10 s and 60 s.

4.5 The graph below shows how the mass of carbon dioxide given off during a chemical reaction varies with time.

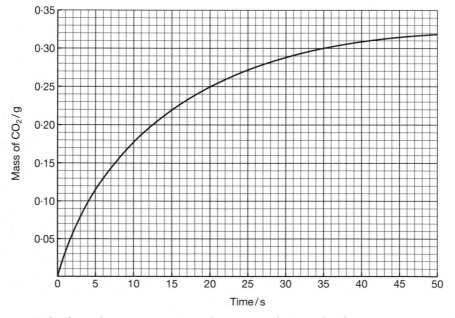

(*a*) Calculate the average rate of reaction during the first 20 s.
(*b*) Calculate the average rate of reaction between 22 s and 42 s.

4.6 The graph below shows how the volume of carbon dioxide produced during a chemical reaction varies with time.

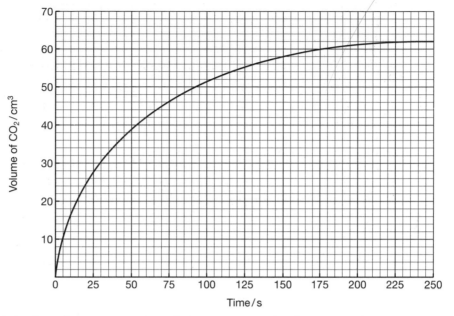

(a) Calculate the average rate of reaction over the first 100 s.
(b) Calculate the average rate of reaction between 125 s and 250 s.

4.7 The graph below shows how the concentration of acid changes during the progress of a chemical reaction.

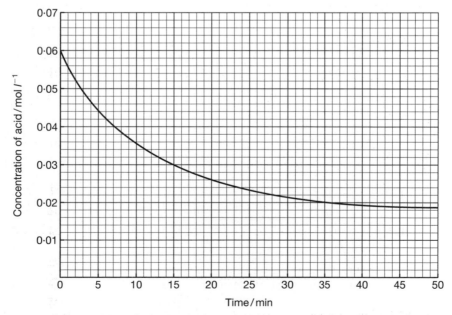

(a) Calculate the average rate of reaction during the first 15 min of the reaction.
(b) Calculate the average rate of reaction between 20 min and 35 min.

4.8 The graph below shows how the mass of a beaker and its contents change as a gas is given off during the course of a chemical reaction.

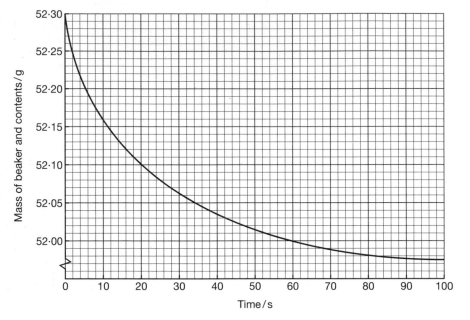

(a) Calculate the average rate of reaction between 20 s and 60 s.

(b) Calculate the average rate of reaction between 22 s and 42 s.

4.9 The graph below shows how the volume of gas produced changes during a chemical reaction.

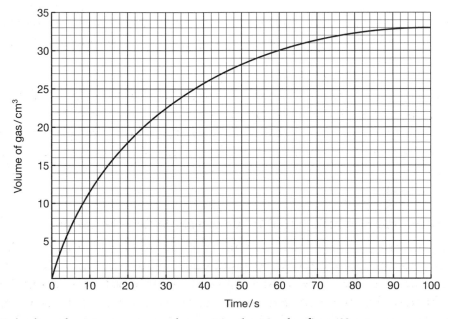

(a) Calculate the average rate of reaction during the first 60 s.

(b) Calculate the average rate of reaction between 50 s and 100 s.

4.10 The graph below shows how the concentration of a reactant changes during the chemical reaction.

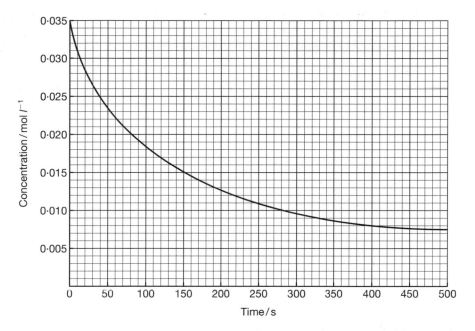

(a) Calculate the average rate of reaction during the first 250 s.

(b) Calculate the average rate of reaction between 80 s and 280 s.

PART 2

In this part of the chapter we consider the interpretation of graphs obtained from 'clock reaction' data. A clock reaction is so called because, when two solutions containing the reactants are mixed, nothing appears to happen for some time and then a quick colour change takes place.

Three common clock reactions are described here. In each, much of the detailed chemistry is omitted, because the way the reactions take place is quite complicated. What is important to note is that there is a colour change which takes place after a period of time. The time taken for the change will be different under different conditions of concentration or temperature and can be used to study the effect of these changes on the rate of reaction.

THE HYDROGEN PEROXIDE/POTASSIUM IODIDE REACTION

Hydrogen peroxide and potassium iodide can react to produce iodine. If the reaction mixture contains sodium thiosulphate and starch solution, as the iodine is produced the thiosulphate immediately reacts with it, converting it back to iodide ions. However, when the thiosulphate is completely used up, the iodine produced can no longer react with it. The moment this 'free' iodine is produced, it reacts with the starch to produce a blue-black colour which appears suddenly. (This should be very familiar from earlier work on carbohydrates where iodine was used to test for the presence of starch, the blue-black colour being a positive result. Here we are using starch as a test for the presence of iodine, exactly the same reaction.)

THE OXALIC ACID/POTASSIUM PERMANGANATE REACTION

The permanganate ion is intensely purple. Oxalic acid can undertake a redox reaction with permanganate where the former is oxidised and the latter is reduced. As the reaction proceeds, the purple colour of the permanganate fades, disappearing completely after a time.

THE SODIUM THIOSULPHATE/HYDROCHLORIC ACID REACTION

When the thiosulphate ion reacts with hydrochloric acid, sulphur is produced. As sulphur does not dissolve properly, it appears as a cloudiness of particles throughout the solution. This can take some time from first appearance to complete cloudiness, so it is difficult to judge the same level of cloudiness between one experiment and another. What is usually done to compare the time for cloudiness to appear in different experiments fairly is to place the beaker on top of a piece of paper with a cross drawn on it. Looking down on the beaker, as the reaction proceeds, the cross starts to disappear from sight. The time taken when the cross *just* disappears can be used to compare the speed of the reaction under different conditions. An important point of procedure is that the shape and size of beaker and the total volume of solution must be the same in each experiment. (It would be much easier to see the cross through a shallow solution, such as a smaller volume, or the same volume in a wider beaker, even with the same level of cloudiness. To compare the 'cloudiness time' fairly, these factors have to be kept the same.

Each of these three 'clock reactions' can be used to investigate the factors which affect the speed or rate of the chemical reaction involved. As the two solutions are mixed, a stop clock is started. When the colour change takes place, or the cloudiness appears, the clock is stopped and the time taken is noted. The same reaction is then repeated under different conditions of temperature, concentration etc.

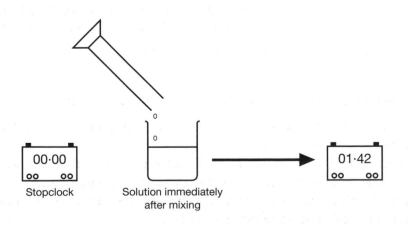

00·00	01·42	
Stopclock	Solution immediately after mixing	Cloudiness or colour change taking place after a certain time

The time taken for a colour change, or a cross to disappear, is an easy quantity to measure and this is what makes clock reactions so handy for studying the factors affecting reaction rates. However, the time taken is not **directly** a measure of the rate of reaction. A moment's thought tells us that the **longer the time taken** for the colour change, the **slower** the reaction. And, of course, the **shorter the time taken** the **faster** the reaction. The relationship between the time taken and the rate of reaction is known in mathematics as **inverse proportion** (or **variation**). This simply means that, as one quantity increases, the other quantity decreases.

We can turn our 'time taken' (t) measurement into a value which gives a measure of rate of reaction simply by taking its **reciprocal** ($1/t$). That is:

$$\text{rate of reaction} = \frac{1}{t}$$

To illustrate this, consider the table below containing the simplified results from a series of experiments in which increasing concentrations of a solution were used in a clock reaction, all other factors being kept constant.

Concentration (arbitrary units)	Time taken t/s	Rate $\frac{1}{t}$/s^{-1}
1	50	0.02
2	25	0.04
3	16.7	0.06
4	12.5	0.08
5	10	0.10

As might be expected, the time taken for the colour change gets shorter as the concentration is increased in each successive experiment from 1 unit to 5 units. That is, the reaction gets faster as we increase the concentration. The right hand column, entitled 'Rate', has $1/t$ calculated for each of the values for t. It can be seen that **these** values go up in line with the increasing concentration; that is, they give a value which is proportional to the rate of reaction.

Before going on to consider graphs of such data, some points need to be noted.

- Firstly, since the unit of 'time taken' will be s or min, the unit of rate, obtained by calculating $1/t$, will be s^{-1} or min^{-1}.
- Secondly, if rate is obtained by taking $1/t$, t can be obtained by calculating $1/\text{rate}$. If this is not immediately obvious, practise converting time to rate and rate to time using the figures in the table. (But note that the figures for concentration = 3 units involved rounding in the calculation of rate.)

- Lastly, most scientific calculators have a button, usually labelled '1/x', which will carry out the reciprocal calculation directly; if not already familiar with its use, it is worth getting practice using it in the examples in this part of the chapter.

Rate of reaction studies involving clock reaction data usually involve carrying experiments out using different concentrations of solutions or at different temperatures. The Worked Examples following illustrate these two situations, although the method used for interpreting the graphs is the same in each case.

Worked Example 4.3

The graph below shows the rate of a clock reaction against concentration of solution. The rate is expressed as the reciprocal of the time taken for cloudiness to appear.

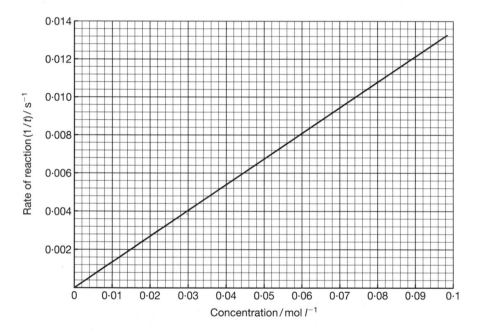

(a) How long did the cloudiness take to appear when the concentration was 0.03 mol l^{-1}?

(b) At what concentration did the cloudiness take 100 s to appear?

(a) From the graph, at a concentration of 0.03 mol l^{-1}, the rate of reaction is 0.004 s^{-1}. The time taken is the **reciprocal** of the rate; that is:

$$\text{time taken} = \frac{1}{0.004}$$

$$= 250 \text{ s}$$

(*b*)   The rate of reaction is the **reciprocal** of the time taken for reaction; that is, when the time taken is 100 s, the rate of reaction is given by:

$$\text{rate} = \frac{1}{100}$$

$$= 0.01 \text{ s}^{-1}$$

From the graph, this value for the rate of reaction occurs at a concentration of 0.074 mol l^{-1}.

Worked Example 4.4

The graph below shows how the rate of a clock reaction varies with the temperature at which it is carried out. The rate is expressed as the reciprocal of the time taken for a colour change to appear.

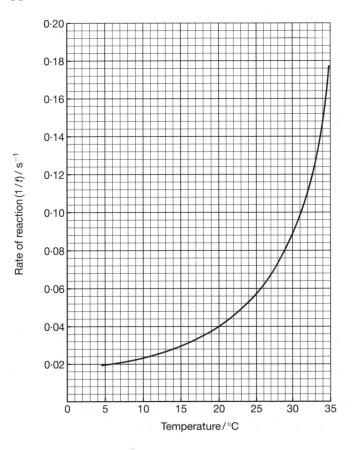

(*a*)   After how long did the colour change take place when the reaction was carried out at 29 °C?

(*b*)   At what temperature would the colour change have taken 25 s to occur?

(*a*)   From the graph, at 29 °C the rate of reaction is 0.08 s^{-1}. The time taken is the reciprocal of the rate. That is:

$$\text{time taken} = \frac{1}{0.08}$$
$$= 12.5 \text{ s}$$

(*b*)   The rate is the reciprocal of time. So, if the reaction took 25 s for the colour change to take place, the rate is expressed by:

$$\text{rate} = \frac{1}{25}$$
$$= 0.04 \text{ s}^{-1}$$

From the graph, the reaction has a rate of 0.04 s^{-1} when the temperature is **20 °C**.

PROBLEMS

Problems 4.11–4.15 are of the type illustrated in Worked Examples 4.3 and 4.4.

4.11   The graph below shows how the rate of a clock reaction varies with the concentration at which it is carried out. The rate is expressed as the reciprocal of the time taken for a colour change to appear.

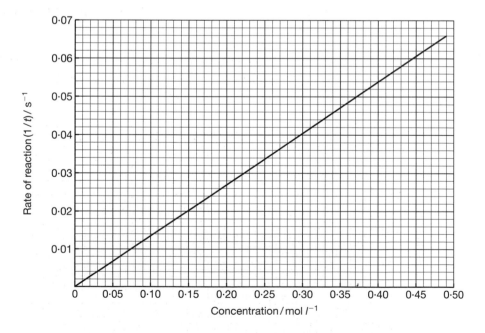

(*a*)   At what concentration does the colour change take 20 s to occur?

(*b*)   After what time will the colour change take place if the reaction is carried out at a concentration of 0.15 mol l^{-1}.

4.12 The graph below shows how the rate of a clock reaction varies with the temperature at which it is carried out. The rate is expressed as the reciprocal of the time taken for cloudiness to appear.

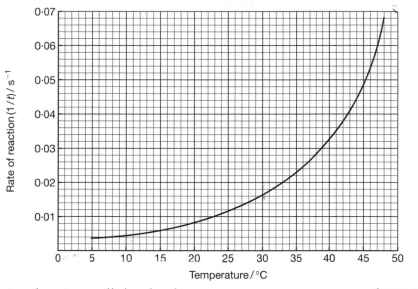

(a) At what time will the cloudiness appear at a temperature of 43 °C?

(b) At what temperature will the cloudiness appear after 50 s?

(c) Obtain from the graph the values for the rate of reaction at 10 °C, 20 °C, 30 °C and 40 °C. What relationship between the temperatures and the rates can be observed?

4.13 The graph below shows how the rate of a clock reaction varies with the concentration at which it is carried out. The rate is expressed as the reciprocal of the time taken for a colour change to appear.

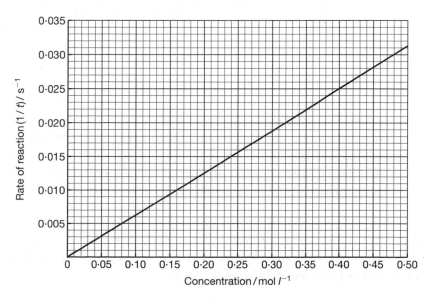

    (*a*)   What concentration of reactant will cause the colour change in this reaction to take place after 50 s?

    (*b*)   After what time will the colour change take place if the reagent concentration is 0.40 mol l^{-1}.

4.14   The graph below shows how the rate of a clock reaction varies with the temperature at which it is carried out. The rate is expressed as the reciprocal of the time taken for a colour change to appear.

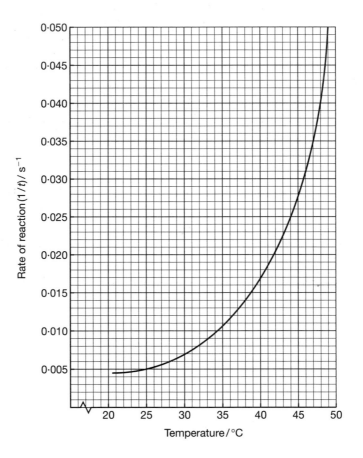

    (*a*)   At what temperature will the colour change take place 200 s after the reagents are mixed?

    (*b*)   At what time after the reagents are mixed will the colour change take place when the temperature of the solution is 48 °C?

4.15   The graph below shows how the rate of a clock reaction varies with the concentration at which it is carried out. The rate is expressed as the reciprocal of the time taken for a colour change to appear.

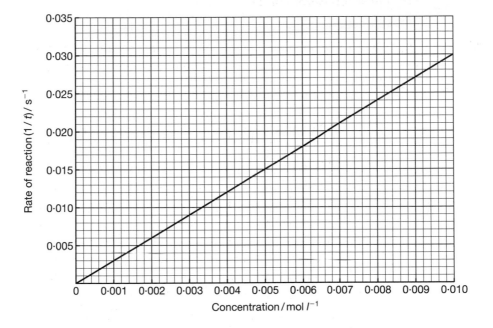

(a)   How long will the colour change take to appear if the reagent concentration is 0.0054 mol l^{-1}.

(b)   At what concentration of reagent will the colour change take 40 s to appear?

5

Calculations from Equations – 3: Involving an Excess of a Reactant

In Chapters two and three involving calculations from equations, a quantity of **one** substance has been given and the quantity of **another** has been asked for. In these problems, it is always stated that **all** of the known substance reacts, usually by saying that the known substance reacts with an **excess** (that is more than enough) of the other reactant.

In this chapter, the quantities of two reactants will be given and the quantity of a product will be asked for. In these examples, one of the reactants will be **in excess**; that is there is **too much** of it to react with all of the other reactant. To calculate the amount of product made, we need to calculate the number of moles of **both** reactants and identify which one is in excess. We then use the number of moles of the **other** reactant as our 'known' moles, since we know that **all** of it will react. The calculation of the quantity of product is carried out exactly as in Chapters two and three.

This type of problem is best seen by following the Worked Example below.

Worked Example 5.1

1.308 g of zinc is reacted with 25 cm^3 of 2 mol l^{-1} hydrochloric acid.

(a)   Calculate which reactant is in excess;
(b)   Calculate the mass of hydrogen gas which will have been given off when the reaction is complete.

Step 1: Balanced Equation

$$Zn + 2HCl \longrightarrow ZnCl_2 + H_2$$

Step 2: Mole Statement
Since we are told the quantities of both the zinc and the hydrochloric acid, and asked for the quantity of hydrogen, we include all three in the Mole Statement:

1 mol of Zn reacts with 2 mol of HCl to produce 1 mol of H$_2$

Step 3: Calculation of 'Known' Moles
We have two known quantities, the Zn and the HCl, so we calculate the number of moles of each, as below.

Zn: 1 mol of Zn $= 65.4$ g

$$65.4 \text{ g} \quad = \quad 1 \quad \text{mol}$$

$$1 \text{ g} \quad = \quad \frac{1}{65.4} \quad \text{mol}$$

$$1.308 \text{ g} = \quad \frac{1.308}{65.4} \quad \text{mol}$$

$$= 0.02 \text{ mol of Zn}$$

HCl: no. of mol $=$ concentration $\times$ no. of litres

$$= \quad 2 \quad \times \quad 0.025$$

$$= 0.05 \text{ mol of HCl}$$

So we have 0.02 mol of Zn and 0.05 mol of HCl. But from the Mole Statement, we see that:

1 mol of Zn will react with 2 mol of HCl.

So 0.02 mol of Zn will react with 0.04 mol of HCl.

The amount of Zn that we actually have is 0.02 mol, which will react with 0.04 mol of HCl. But there is 0.05 mol of HCl present, which is more than will react; **the HCl is therefore in excess.**

This is the answer to part *(a)* of the problem.

In the final stages of the problem, we can use either 0.02 mol of Zn or 0.04 mol of HCl as the 'known' substances as these amounts will completely react with each other. We **cannot** use the value of 0.05 mol of HCl, since not all of it will react. (0.04 mol will, leaving an unreacted excess of 0.01 mol.)

Step 4: Calculation of 'Unknown' Moles

1 mol of Zn reacts with 2 mol of HCl to produce 1 mol of H_2

0.02 mol of Zn reacts with 0.04 mol of HCl to produce **0.02 mol of H_2**

Step 5: Finishing Off
We have calculated that 0.02 mol of H_2 will be produced and we have to express this as a mass:

$$1 \quad \text{mol of } H_2 = \quad 2 \text{ g}$$

$$0.02 \text{ mol of } H_2 = \quad 0.02 \times 2 \text{ g}$$

$$= 0.04 \text{ g}$$

PROBLEMS

These problems are of the type illustrated in Worked Example 5.1, in which one of the reactants is in excess.

5.1
$$Pb(s) + 2HCl(aq) \longrightarrow PbCl_2(aq) + H_2(g)$$

6.216 g of lead is added to 50 cm³ of 1 mol l^{-1} hydrochloric acid.

(a)   Calculate which reactant is in excess;

(b)   What mass of hydrogen gas will have been given off on completion of the reaction?

5.2 $CaCO_3(s) + 2HNO_3(aq) \longrightarrow Ca(NO_3)_2(aq) + CO_2(g) + H_2O(l)$

12 g of calcium carbonate is placed in 500 cm³ of 0.4 mol l^{-1} nitric acid.

(a)   Calculate which reactant is in excess;

(b)   What mass of carbon dioxide would have been given off when the reaction was complete?

5.3
$$Pb(NO_3)_2(aq) + 2KI(aq) \longrightarrow PbI_2(s) + 2KNO_3(aq)$$

120 cm³ of 0.2 mol l^{-1} lead(II) nitrate solution is added to 200 cm³ of 0.25 mol l^{-1} potassium iodide solution. The lead(II) iodide precipitate formed is filtered and dried.

(a)   Calculate which reactant is in excess;

(b)   What is the theoretical mass of precipitate obtained?

5.4 $2CH_3COOH(l) + Na_2CO_3(aq) \longrightarrow 2CH_3COONa(aq) + CO_2(g) + H_2O(l)$

3 g of pure ethanoic acid, CH_3COOH, is added to 90 cm³ of 0.4 mol l^{-1} sodium carbonate solution.

(a)   Calculate which reactant is in excess;

(b)   What mass of carbon dioxide will be evolved?

5.5 $Al(NO_3)_3(aq) + 3NaOH(aq) \longrightarrow Al(OH)_3(s) + 3NaNO_3(aq)$

40 cm³ of 0.5 mol l^{-1} aluminium nitrate solution is added to 200 cm³ of 0.4 mol l^{-1} sodium hydroxide solution.

(a)   Calculate which reactant is in excess;

(b)   What mass of aluminium hydroxide would be precipitated?

5.6 $2AgNO_3(aq) + MgCl_2(aq) \longrightarrow 2AgCl(s) + Mg(NO_3)_2(aq)$

40 cm³ of 0.045 mol l^{-1} silver(I) nitrate solution is added to 12 cm³ of 0.1 mol l^{-1} magnesium chloride solution.

(a)   Calculate which reactant is in excess;

(b)   What mass of silver(I) chloride would be precipitated?

5.7 $$H_2SO_4(aq) + BaCl_2(aq) \longrightarrow BaSO_4(s) + 2HCl(aq)$$

$20\ cm^3$ of $0.2\ mol\ l^{-1}$ sulphuric acid is added to $50\ cm^3$ of $0.1\ mol\ l^{-1}$ barium chloride solution.

(*a*)   Calculate which reactant is in excess;

(*b*)   What mass of barium sulphate would be precipitated?

5.8 $$(NH_4)_2SO_4(aq) + 2NaOH(aq) \longrightarrow Na_2SO_4(aq) + 2H_2O(l) + 2NH_3(g)$$

$6.605\ g$ of ammonium sulphate is added to $120\ cm^3$ of $0.5\ mol\ l^{-1}$ sodium hydroxide solution and the solution heated.

(*a*)   Calculate which reactant is in excess;

(*b*)   What is the theoretical mass of ammonia which would be evolved?

5.9 $$FeS(s) + 2HCl(aq) \longrightarrow FeCl_2(aq) + H_2S(g)$$

$4.395\ g$ of iron(II) sulphide is added to $160\ cm^3$ of $0.5\ mol\ l^{-1}$ hydrochloric acid.

(*a*)   Calculate which reactant is in excess;

(*b*)   What is the maximum mass of hydrogen sulphide gas, H_2S, which will be given off?

5.10 $$CuO(s) + H_2SO_4(aq) \longrightarrow CuSO_4(aq) + H_2O(l)$$

$1.59\ g$ of copper(II) oxide is added to a beaker of $50\ cm^3$ of $0.24\ mol\ l^{-1}$ sulphuric acid. The mixture is heated and stirred until no further reaction takes place, and the contents of the beaker are filtered.

(*a*)   Calculate which reactant is in excess;

(*b*)   What mass of unreacted copper(II) oxide would be removed from the beaker by the filtration, after being dried?

(*c*)   The remaining solution is evaporated and dried to form anhydrous copper(II) sulphate, $CuSO_4$. What mass of copper(II) sulphate would be obtained?

6

Enthalpy Changes (ΔH)

The enthalpy change, ΔH, for a chemical reaction is expressed as the amount of heat energy (in kJ) given out or taken in for every mole of a particular substance reacted or product formed.

We can measure the heat given out or taken in during certain reactions by using it to increase or decrease the temperature of a known mass of water. We can then use the equation below to calculate the amount of heat involved:

$$\Delta H = cm\Delta T$$

- ΔH is the heat given out or taken in during the reaction. This is measured in kilojoules (kJ). If heat is given out (an exothermic reaction), a **negative** sign is given to the numerical value. For example, a ΔH value of -40 kJ would mean that 40 kJ of heat had been given out. If heat is taken in (an endothermic reaction), the numerical value is **positive**. Although it is not strictly necessary to put a $+$ sign before the value, in this book this will be done to emphasise the difference between exothermic and endothermic reactions.

  When the heat being given out (or taken in) refers to 1 mol of substance reacting, the ΔH is expressed in kJ mol^{-1}.

- c is a constant, called the specific heat capacity, for the substance being heated or cooled. In all the examples we will consider, this substance will be water (or mainly water); the value of c for water is 4.18 kJ kg^{-1} °C^{-1}. This value refers to the fact that if 4.18 kJ of heat is added to (or taken out of) 1 kg of water, the temperature will rise (or fall) by 1 °C.

  In the dissolving or neutralisation processes, it is not **pure** water, but a **solution** whose temperature is being changed. Can we use the value of 4.18 kJ kg^{-1} °C^{-1} in these cases? Strictly speaking, the answer is no, but the solutions involved are dilute and we can assume that the error involved in using this value is very small in comparison to the errors caused by loss or gain of heat by the solution during the experiment. In this chapter, we will use the value of $c = 4.18 \, \text{kJ} \, \text{kg}^{-1} \, °\text{C}^{-1}$ in **all** cases where water or an aqueous solution is heating up or cooling down.

- m is the mass of water (or aqueous solution) being heated (or cooled down), expressed in kg. Often, the amount of water is expressed as a volume, in cm^3 or litres. The conversion of the volume of water to a mass is simple because $1 \, \text{cm}^3$ of water weighs $1 \, \text{g}$.

- ΔT is the temperature change in °C. This is expressed as a positive value, whether or not the change is an increase or decrease in temperature. As noted in the first bullet point above, the sign of the ΔH value (+ or −) will indicate whether heat has been taken in or given out during the reaction.

PART A THE ENTHALPY OF SOLUTION AND THE ENTHALPY OF COMBUSTION

Three enthalpy changes which are measured experimentally as part of the Higher syllabus are the **Enthalpy of Combustion**, the **Enthalpy of Solution** and the **Enthalpy of Neutralisation**. The first two of these are considered together in this part of the Chapter because the calculations involved are very similar; the Enthalpy of Neutralisation is considered later, in Part B.

The Enthalpy of Solution
The Enthalpy of Solution is the amount of heat which is given out (or taken in) when 1 mol of a substance is dissolved completely in water. Some dissolving processes are **exothermic,** meaning that heat is given **out,** causing the temperature to rise. Other processes are **endothermic,** where heat is taken **in,** causing the temperature to fall.

The Enthalpy of Solution of a substance can be measured experimentally by weighing an amount of it and dissolving it in a known mass of water. The temperature of the water is taken before and after complete dissolving. The quantity of heat given out (or taken in) can be calculated using the equation $\Delta H = cm\Delta T$. The amount of heat involved, had **1 mol** of the substance been dissolved, can then be calculated by simple proportion, as before, knowing its formula, and therefore, the mass of 1 mol.

The Enthalpy of Combustion
The Enthalpy of Combustion is the amount of heat given out when 1 mol of a specified substance is burned completely in an excess of oxygen.

The Enthalpy of Combustion of a flammable liquid, such as an alcohol, can be measured using the apparatus shown below.

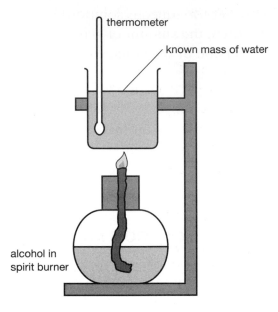

A known mass of water is measured into a metal can and the temperature of the water is taken. The burner containing alcohol is weighed, lit and placed under the can. After some time, when the water has heated up sufficiently, the flame is put out and the burner is reweighed. The highest temperature that the water reaches is then taken. (It takes some time for all the heat in the metal can to get to the water.)

The heat produced by the burning can then be calculated using the equation $\Delta H = cm\Delta T$. The mass of alcohol burned is obtained by subtracting the final mass of the burner from its initial mass. The amount of heat which would have been produced had **1 mol** of the alcohol been burned can then be calculated by simple proportion, knowing its formula, and therefore, the mass of 1 mol.

It should be noted that this is a very inaccurate method of measuring the Enthalpy of Combustion of a substance because so much heat is lost to the surroundings. Data Book values are obtained by much more accurate methods which reduce heat loss to a minimum.

The use of this equation $\Delta H = cm\Delta T$ in calculating quantities of heat can be seen in the following Worked Examples.

Worked Example 6.1

A substance is dissolved in 0.3 kg of water, causing a temperature rise of 5 °C. What quantity of heat has been given out?

In this problem,

$$m = 0.3\,\text{kg}$$
$$c = 4.18\,\text{kJ}\,\text{kg}^{-1}\,{}^{\circ}\text{C}^{-1}$$
$$\Delta T = 5\,{}^{\circ}\text{C}$$

We fit these figures into the equation shown below:

$$\Delta H = cm\Delta T$$
$$= 4.18 \times 0.3 \times 5$$
$$= 6.27\,\text{kJ}$$

This calculated value is the quantity of heat given out in the dissolving process described. Since heat has been given **out**, the enthalpy change would be given a negative sign, ie, −6.27 kJ.

Worked Example 6.2

0.02 mol of an alcohol is burned in a spirit burner. The heat given out is used to raise the temperature of 0.4 kg of water by 10 °C. Calculate the Enthalpy of Combustion of the alcohol.

The Enthalpy of Combustion of a compound is the quantity of heat given out when **1 mol** of the compound is burned completely.

Firstly, we calculate the quantity of heat produced:

$$\Delta H = cm\Delta T$$
$$= 4.18 \times 0.4 \times 10$$
$$= 16.72\,\text{kJ}$$

This is the amount of heat given out when **0.02 mol** of the compound was burned. We now calculate the amount of heat which would be given out if **1 mol** had been burned.

0.02 mol burns giving out 16.72 kJ of heat

1 mol burns giving out $\dfrac{16.72}{0.02}$ kJ of heat

$$= 836\,\text{kJ}$$

So the Enthalpy of Combustion of the substance is −836 kJ mol⁻¹.

Note that the negative sign is included to show that this is heat given out; that is, it is an exothermic reaction. Be aware, however, that some text books and data books do not include the negative sign before combustion enthalpies. (Since combustions are **always** exothermic, it is argued that the sign is therefore not necessary, except when carrying out the kind of calculations shown in Chapter 12.) In this book, however, the signs showing the direction of enthalpy changes will **always** be shown.

Worked Example 6.3

0.3 mol of a salt is dissolved in 0.25 kg of water, causing a rise in the temperature of the solution from 17 °C to 23 °C. Calculate the Enthalpy of Solution of the salt.

The Enthalpy of Solution is the quantity of heat given out or taken in when 1 mol of a particular substance is dissolved completely in water.

Firstly, we calculate the quantity of heat produced, remembering that ΔT is the change in temperature which, in this example, is 6 °C.

$$\Delta H = cm\Delta T$$
$$= 4.18 \times 0.25 \times 6$$
$$= 6.27\,kJ$$

This is the amount of heat given out when 0.3 mol of the salt dissolves.

We now calculate the amount of heat which would be given out if **1 mol** were dissolved.

0.3 mol dissolves, giving out 6.27 kJ of heat

1 mol dissolves, giving out $\dfrac{6.27}{0.3}$ kJ of heat

$$= 20.9\ kJ$$

So the Enthalpy of Solution of the salt is $-20.9\,kJ\,mol^{-1}$.

Worked Example 6.4

0.16 g of methanol, CH_3OH, is burned in a spirit burner. The heat from this combustion causes the temperature of 0.1 kg of water to be raised from 20 °C to 27 °C. Use this information to calculate the Enthalpy of Combustion of methanol.

Fitting the data into the equation, we have:

$$\Delta H = cm\Delta T$$
$$= 4.18 \times 0.1 \times 7$$
$$= 2.926\,kJ$$

The calculated value of 2.926 kJ is the quantity of heat which is given out when 0.16 g of methanol is burned.

We now calculate the heat which would have been given out if **1 mol** of methanol had been burned.

The formula of methanol is CH_3OH. So 1 mol of methanol = 32 g.

0.16 g of methanol burns giving out 2.926   kJ of heat

1   g of methanol burns giving out $\dfrac{2.926}{0.16}$ kJ of heat

32   g of methanol burns giving out $\dfrac{2.926 \times 32}{0.16}$ kJ of heat

$$= 585.2\,kJ$$

Thus, the calculated Enthalpy of Combustion of methanol is $-585\,kJ\,mol^{-1}$ (rounded to three significant figures).

Worked Example 6.5

4 g of ammonium nitrate, NH_4NO_3, is dissolved completely in 100 cm^3 of water in an insulated container. The temperature of the water falls from 19 °C to 16 °C. Calculate the Enthalpy of Solution of ammonium nitrate.

Note that in this problem, the **volume** of water is given rather than the **mass**. Since it is known that 1 litre (1000 cm^3) of water weighs 1 kg (1000 g), the conversion of 100 cm^3 to 0.1 kg is straightforward.

The data are now put into the equation:

$$\Delta H = cm\Delta T$$
$$= 4.18 \times 0.1 \times 3$$
$$= 1.254\,kJ$$

The calculated value of 1.254 kJ is the quantity of heat taken in by 4 g of ammonium nitrate from the water. We now need to obtain the quantity of heat which would be taken in by **1 mol** of ammonium nitrate.

The formula of ammonium nitrate is NH_4NO_3. So 1 mol of NH_4NO_3 = 80 g.

4 g of ammonium nitrate dissolves, taking in 1.254   kJ of heat

1 g of ammonium nitrate dissolves, taking in $\dfrac{1.254}{4.0}$ kJ of heat

80 g of ammonium nitrate dissolves, taking in $\dfrac{1.254 \times 80}{4.0}$ kJ of heat

$$=\quad 25.08\,kJ$$

So the calculated Enthalpy of Solution of ammonium nitrate is $+25.1\,kJ\,mol^{-1}$.

PROBLEMS

- Problems 6.1 to 6.5 involve simple calculations of the type shown in Worked Example 6.1.
- Problems 6.6 to 6.10 involve calculations of the Enthalpies of Solution and Combustion of the type shown in Worked Examples 6.2 and 6.3.
- Problems 6.11 to 6.20 involve calculations of the Enthalpies of Solution and Combustion of the type shown in Worked Examples 6.4 and 6.5.

6.1 A substance is dissolved in 0.2 kg of water, causing a temperature rise of 4 °C. What quantity of heat has been given out?

6.2 A certain mass of alcohol is burned in a spirit burner which is used to heat up 0.1 kg of water in a metal can. The temperature of the water rises by 8 °C. What quantity of heat has been given out?

6.3 A 2 °C fall in temperature is recorded when a substance is dissolved in 0.25 kg of water. What quantity of heat has been absorbed by the water?

6.4 When a quantity of a substance is dissolved in 0.2 kg of water, the temperature rises from 19.5 °C to 21 °C. What quantity of heat has been given out?

6.5 A Bunsen burner is used to heat 0.5 kg of water from 20.5 °C to 39.5 °C. How much heat has been produced in the burning of the gas?

6.6 0.01 mol of a substance is dissolved in 0.2 kg of water, causing the temperature to drop from 19 °C to 17 °C. Calculate the Enthalpy of Solution of the substance.

6.7 0.02 mol of a hydrocarbon is burned completely in air and the heat produced is used to heat 0.1 kg of water from 20.5 °C to 29.5 °C. Calculate the Enthalpy of Combustion of the fuel.

6.8 0.025 mol of a salt is dissolved in 0.4 kg of water at 20 °C. The temperature of the solution rises to 25 °C. Calculate the Enthalpy of Solution of the salt.

6.9 A can containing 0.1 kg of water at 21 °C is heated to 29 °C when a burner containing an alcohol is lit underneath it. If 0.02 mol of the alcohol is burned in the process, calculate its Enthalpy of Combustion.

6.10 0.05 mol of a compound is dissolved in 0.5 kg of water, causing its temperature to fall from 21 °C to 19.5 °C. Calculate the Enthalpy of Solution of the compound.

6.11   0.32 g of methanol, CH_3OH, is burned in a spirit burner which is used to heat up 0.2 kg of water from 19.5 °C to 27.5 °C. Calculate the Enthalpy of Combustion of methanol.

6.12   5.3 g of sodium carbonate, Na_2CO_3, is dissolved in 0.1 kg of water, causing the temperature to rise from 20.5 °C to 23.5 °C. Calculate the Enthalpy of Solution of the compound.

6.13   The burning of 0.2 g of methane, CH_4, is used to raise the temperature of 0.25 kg of water from 18.5 °C to 28.5 °C. Calculate the Enthalpy of Combustion of methane.

6.14   8 g of ammonium nitrate, NH_4NO_3, is dissolved in 0.2 kg of water, causing the temperature to drop from 20 °C to 17 °C. Calculate the Enthalpy of Solution of ammonium nitrate.

6.15   A burner containing ethanol, C_2H_5OH, is used to heat up 0.4 kg of water from 21 °C to 37 °C. In the process, 0.92 g of ethanol is burned. Calculate the Enthalpy of Combustion of ethanol.

6.16   4 g of sodium hydroxide, NaOH, is dissolved in 0.25 kg of water, causing the temperature to rise from 19 °C to 23 °C. Calculate the Enthalpy of Solution of sodium hydroxide.

6.17   0.22 g of propane, C_3H_8, is burned to heat 0.25 kg of water from 21 °C to 31 °C. Calculate the Enthalpy of Combustion of propane.

6.18   14.92 g of potassium chloride, KCl, is dissolved in 0.2 kg of water. The temperature falls from 19.5 °C to 15.5 °C. Calculate the Enthalpy of Solution of the salt.

6.19   A gas burner containing butane, C_4H_{10}, is used to heat 0.15 kg of water from 22.5 °C to 31 °C. 0.116 g of butane is burned in the process. Calculate the Enthalpy of Combustion of butane.

6.20   2.022 g of potassium nitrate, KNO_3, is dissolved in 0.1 kg of water, causing the temperature to fall from 20.5 °C to 18.9 °C. Calculate the Enthalpy of Solution of potassium nitrate.

Worked Examples 6.6 and 6.7

The next two Worked Examples involve using known values for the Enthalpy of Solution and Enthalpy of Combustion to calculate other values. These are slightly more complex and need more arithmetical manipulation than the previous calculations in this chapter.

Worked Example 6.6

The Enthalpy of Solution of sodium carbonate, Na_2CO_3, is $-24.7\,kJ\,mol^{-1}$. What mass of sodium carbonate dissolved in $150\,cm^3$ of water would cause a temperature rise of 0.5 °C?

We firstly calculate the amount of heat given out:

$$\Delta H = cm\Delta T$$
$$= 4.18 \times 0.15 \times 0.5$$
$$= 0.3135\,kJ$$

Next, we note that:

The Enthalpy of Solution of Na_2CO_3 is $-24.7\,kJ\,mol^{-1}$ and 1 mol of $Na_2CO_3 = 106\,g$.

So

24.7 kJ is the heat given out when 106 g of Na_2CO_3 dissolves

1 kJ is the heat given out when $\dfrac{106}{24.7}$ g of Na_2CO_3 dissolves

0.3135 kJ is the heat given out when $\dfrac{106 \times 0.3135}{24.7}$ g of Na_2CO_3 dissolves

$$= 1.35\,g \text{ (after rounding)}$$

Worked Example 6.7

The Enthalpy of Combustion of methanol is $-727\,kJ\,mol^{-1}$. A burner containing methanol, CH_3OH, is used to heat up $400\,cm^3$ of water. What temperature rise would be produced in the water if 0.64 g of methanol were completely burned?

In this problem, we cannot use the equation $\Delta H = cm\Delta T$ immediately, because we are asked for the temperature rise, ΔT, but we do not yet know ΔH – the heat given out in this particular experiment. We firstly have to calculate ΔH from the Enthalpy of Combustion value given.

The Enthalpy of Combustion of $CH_3OH = -727\,kJ\,mol^{-1}$ and 1 mol of $CH_3OH = 32\,g$

So we can calculate the amount of heat given out (ΔH) when 0.64 g burns as follows:

32 g of CH_3OH burns giving out 727 kJ of heat

1 g of CH_3OH burns giving out $\dfrac{727}{32}$ kJ of heat

0.64 g of CH_3OH burns giving out $\dfrac{727 \times 0.64}{32}$ kJ of heat

$$= 14.54\,kJ$$

We can now put this value for ΔH, and the other known values, into the equation $\Delta H = cm\Delta T$, after rearranging it to put ΔT on the left hand side.

$$\Delta H = cm\Delta T$$

So:

$$\Delta T = \frac{\Delta H}{cm}$$

$$= \frac{14.54}{4.18 \times 0.4}$$

$$= 8.7\,°C \text{ (after rounding)}$$

PROBLEMS

Problems 6.21–6.30 are of the type shown in Worked Examples 6.6 and 6.7, using values for Enthalpies of Solution and Combustion given in the questions.

6.21 The Enthalpy of Solution of sodium hydroxide, NaOH, is $-42.6\,kJ\,mol^{-1}$. What mass of sodium hydroxide would produce a temperature rise of 5 °C when dissolved completely in 200 cm³ of water.

6.22 The Enthalpy of Combustion of ethanol, C_2H_5OH, is $-1367\,kJ\,mol^{-1}$. If a spirit burner containing ethanol was used to heat a can of water, what mass of ethanol would raise the temperature of 300 cm³ of water by 10 °C?

6.23 The Enthalpy of Solution of sodium carbonate, Na_2CO_3, is $-24.7\,kJ\,mol^{-1}$. Calculate the temperature change which would take place if 1.325 g of sodium carbonate were dissolved completely in 250 cm³ of water.

6.24 The Enthalpy of Combustion of propanol is $-2020\,kJ\,mol^{-1}$. A burner containing propanol, C_3H_7OH, is used to heat up 200 cm³ of water. What mass of propanol would require to be burned to produce a temperature rise of 13.5 °C?

6.25 The Enthalpy of Solution of potassium hydroxide, KOH, is $-55.2\,kJ\,mol^{-1}$. What temperature change would take place if 2.805 g of potassium hydroxide were dissolved completely in 100 cm³ of water?

6.26 A Bunsen burner uses methane, CH_4, which has an Enthalpy of Combustion of $-891\,kJ\,mol^{-1}$. If 0.4 g of methane were completely burned to heat a can containing 500 cm³ of water, what would be the maximum temperature rise which would be produced?

6.27 The Enthalpy of Solution of ammonium chloride, NH_4Cl, is $+15.0\,kJ\,mol^{-1}$. What mass of ammonium chloride would require to be dissolved in 200 cm³ of water to lower the temperature by 2 °C?

6.28   The Enthalpy of Combustion of propane is $-2220\,\text{kJ}\,\text{mol}^{-1}$. What would be the minimum mass of propane, C_3H_8, which would need to be completely burned in a propane burner to bring 5 kg of water to the boil from an initial temperature of 20 °C?

6.29   The Enthalpy of Solution of barium hydroxide, $Ba(OH)_2$, is $-51.8\,\text{kJ}\,\text{mol}^{-1}$. What mass of barium hydroxide dissolved in 200 cm^3 of water would cause a temperature rise of 4 °C?

6.30   A camping stove runs on butane, C_4H_{10}. If the Enthalpy of Combustion of butane is $-2877\,\text{kJ}\,\text{mol}^{-1}$, what mass of butane would require to be burned to bring 2 kg of water to the boil from an initial temperature of 20 °C?

PART B: THE ENTHALPY OF NEUTRALISATION

The Enthalpy of Neutralisation is defined as the amount of heat given out when 1 mol of water is produced during the reaction of a strong acid and a strong alkali.

Note that the meaning of the term 'strong' applied to an acid or alkali will not be covered in the Higher course until Unit 3. It is enough at this stage to be aware that the term refers to **types** of acids and alkalis which are fully split up into ions in solution. Strong acids include hydrochloric, nitric and sulphuric; strong alkalis include sodium hydroxide and potassium hydroxide. Other types of acids and alkalis, described as 'weak' will not be referred to in this chapter.

The Enthalpy of Neutralisation has a constant value of $-57.3\,\text{kJ}\,\text{mol}^{-1}$ which applies for **all** strong acid – strong alkali neutralisations. It might seem surprising that the same value applies for apparently different reactions; however the reactions are **not** different. Consider the two neutralisation equations below:

$$HCl(aq) \;+\; NaOH(aq) \longrightarrow NaCl(aq) \;+\; H_2O(l) \quad \Delta H = -57.3\,\text{kJ}\,\text{mol}^{-1}$$

$$HNO_3(aq) \;+\; KOH(aq) \longrightarrow KNO_3(aq) \;+\; H_2O(l) \quad \Delta H = -57.3\,\text{kJ}\,\text{mol}^{-1}$$

The former represents the neutralisation of hydrochloric acid with the alkali, sodium hydroxide; the latter represents the neutralisation of nitric acid with the alkali, potassium hydroxide. But the only reaction which is taking place in both situations is that between the $H^+(aq)$ from the acid and the $OH^-(aq)$ from the alkali to make water. This is shown by the equation below.

$$H^+(aq) \;+\; OH^-(aq) \longrightarrow H_2O(l) \qquad\qquad \Delta H = -57.3\,\text{kJ}\,\text{mol}^{-1}$$

All the other ions present are spectator ions; that is that they are present in the solution, but are not actually taking part in the reaction. Since in any strong acid–strong alkali neutralisation, this same reaction takes place, the value for the Enthalpy of Neutralisation is the same.

The Enthalpy of Neutralisation can be measured by mixing separate solutions of an acid and an alkali, each containing the same number moles of hydrogen ions and hydroxide ions respectively. The temperature of the solutions would be taken before and after mixing; ΔT is obtained by the temperature rise. (If the two solutions are not at the same starting temperature, an average of the two temperatures must be taken. The mass of solution, m, is the **total mass of the mixed solutions**. The heat produced can then be calculated using the equation $\Delta H = cm\Delta T$. The heat which would have been produced had **1 mol** of water been made in the neutralisation can be calculated from knowing the initial number of moles of hydrogen and hydroxide ions present in the acid and alkali respectively.

An example of this type of calculation is given in the following Worked Example.

Worked Example 6.8

$100 \, cm^3$ of $1 \, mol \, l^{-1}$ hydrochloric acid solution and $100 \, cm^3$ of $1 \, mol \, l^{-1}$ sodium hydroxide solution (both at 20 °C) are mixed in an insulated container. The temperature of the solution rises to 27 °C. Calculate the Enthalpy of Neutralisation from this information.

When mixed, the total volume of solution will be $200 \, cm^3$ which has a mass of 0.2 kg. This, and the other data can be put into the equation:

$$\Delta H = cm\Delta T$$
$$= 4.18 \times 0.2 \times 7$$
$$= 5.852 \, kJ$$

The Enthalpy of Neutralisation is defined as the quantity of heat given out when **1 mol** of water is formed in the neutralisation of a strong acid and a strong alkali. What we have calculated is the amount of heat given out when 0.1 mol of HCl neutralises 0.1 mol of NaOH. This results in the formation of 0.1 mol of water as can be seen from the balanced chemical equation below:

$$NaOH \quad + \quad HCl \quad \longrightarrow \quad NaCl \quad + \quad H_2O$$

	1 mol	+	1 mol	1 mol	+	1 mol
So	0.1 mol	+	0.1 mol	0.1 mol	+	0.1 mol

0.1 mol of water is formed, giving out 5.852 kJ of heat

1 mol of water is formed, giving out $\dfrac{5.852}{0.1}$ kJ of heat

$$= 58.52 \, kJ$$

That is, the calculated Enthalpy of Neutralisation is $-58.5 \, kJ \, mol^{-1}$ (after rounding).

PROBLEMS

Problems 6.31–6.35 are of the type shown in Worked Example 6.8, involving the calculation of the Enthalpy of Neutralisation.

6.31 $100\,cm^3$ of $0.5\,mol\,l^{-1}$ nitric acid, HNO_3, and $100\,cm^3$ of $0.5\,mol\,l^{-1}$ potassium hydroxide solution, KOH (both solutions at the same temperature) are mixed in an insulated container. A temperature rise of $3.5\,°C$ is noted. Calculate the Enthalpy of Neutralisation.

6.32 $50\,cm^3$ of $1\,mol\,l^{-1}$ hydrochloric acid, HCl, and $50\,cm^3$ of $1\,mol\,l^{-1}$ potassium hydroxide solution, KOH, both at $20\,°C$, are mixed. The temperature of the resulting solution rises to $26.9\,°C$. Calculate the Enthalpy of Neutralisation.

6.33 $40\,cm^3$ of $1\,mol\,l^{-1}$ nitric acid, HNO_3, and $40\,cm^3$ of $1\,mol\,l^{-1}$ sodium hydroxide, NaOH, are allowed to reach room temperature of $19\,°C$. When the solutions are mixed, the temperature rises to $25.8\,°C$. Calculate the Enthalpy of Neutralisation.

6.34 $80\,cm^3$ of $0.5\,mol\,l^{-1}$ potassium hydroxide solution, KOH, and $80\,cm^3$ of $0.5\,mol\,l^{-1}$ hydrochloric acid, HCl, both at $18.5\,°C$, are mixed. The temperature of the resulting solution rises to $21.9\,°C$. Calculate the Enthalpy of Neutralisation.

6.35 $25\,cm^3$ of $1\,mol\,l^{-1}$ sulphuric acid, H_2SO_4, is neutralised by $50\,cm^3$ of $1\,mol\,l^{-1}$ sodium hydroxide solution, NaOH. A temperature rise of $9.1\,°C$ is noted. Calculate the Enthalpy of Neutralisation.

Worked Example 6.9

$50\,cm^3$ of $0.2\,mol\,l^{-1}$ sodium hydroxide solution, NaOH, is neutralised by $50\,cm^3$ of $0.2\,mol\,l^{-1}$ hydrochloric acid, HCl. Calculate the resulting temperature rise. (Take the Enthalpy of Neutralisation to be $-57.3\,kJ\,mol^{-1}$.)

We firstly calculate how many moles of water have been formed in the neutralisation.

$$\text{No. of moles of NaOH} = 0.05 \times 0.2 = 0.01\,mol$$
$$\text{No. of moles of HCl} = 0.05 \times 0.2 = 0.01\,mol$$

	HCl	+	NaOH	$\longrightarrow$	NaCl	+	H_2O
	1 mol	+	1 mol		1 mol	+	1 mol
So:	0.01 mol	+	0.01 mol		0.01 mol	+	0.01 mol

So 0.01 mol of water has been formed in this neutralisation. We can then use the value given for the Enthalpy of Neutralisation to calculate how much heat (ΔH) would have been given out in this particular case.

1 mol of water is formed, giving out 57.3 kJ of heat

0.01 mol of water is formed, giving out 0.01×57.3 kJ of heat
$$= 0.573 \text{ kJ}$$

We can now put this value for ΔH, and the other known values, into the equation $\Delta H = cm\Delta T$, after rearranging it to put ΔT on the left hand side.

$$\Delta H = cm\Delta T$$

So
$$\Delta T = \frac{\Delta H}{cm}$$

$$= \frac{0.573}{4.18 \times 0.1}$$

$$= 1.37\,°C \text{ (after rounding)}$$

PROBLEMS

Problems 6.36–6.40 are of the type illustrated in Worked Example 6.9, in which the value of the Enthalpy of Neutralisation is used to calculate a temperature rise in a neutralisation reaction.

Note: In these problems take the Enthalpy of Neutralisation of a strong acid with a strong alkali to be -57.3 kJ mol^{-1}.

6.36 60 cm^3 of 0.5 mol l^{-1} potassium hydroxide solution, KOH, is neutralised by 60 cm^3 of 0.5 mol l^{-1} hydrochloric acid, HCl. Calculate the temperature rise which would take place. (Assume both solutions were at the same temperature before mixing.)

6.37 30 cm^3 of 1 mol l^{-1} sodium hydroxide solution, NaOH, is neutralised by 30 cm^3 of 1 mol l^{-1} nitric acid, HNO$_3$. Calculate the resulting temperature rise. (Assume both solutions were at the same temperature before mixing.)

6.38 40 cm^3 of 0.5 mol l^{-1} potassium hydroxide solution, KOH, is neutralised by 40 cm^3 of 0.5 mol l^{-1} hydrochloric acid, HCl. Calculate the resulting temperature rise. (Assume both solutions were at the same temperature before mixing.)

6.39   80 cm³ of 0.5 mol l^{-1} potassium hydroxide solution, KOH, is mixed with 80 cm³ of 0.5 mol l^{-1} nitric acid, HNO_3. Calculate the resulting temperature rise. (Assume both solutions were at the same temperature before mixing.)

6.40   40 cm³ of 1 mol l^{-1} sodium hydroxide solution, NaOH, is mixed with 20 cm³ of 1 mol l^{-1} sulphuric acid, H_2SO_4. Calculate the temperature rise which would take place. (Assume both solutions were at the same tempeature before mixing.)

The Avogadro Constant and the Mole

The term 'mole' meaning the formula mass of a substance, expressed in grams, should already be familiar. In this section, we consider the mole as an actual **number** of particles – called the Avogadro Constant.

Consider 1 mol of each of the following elements:

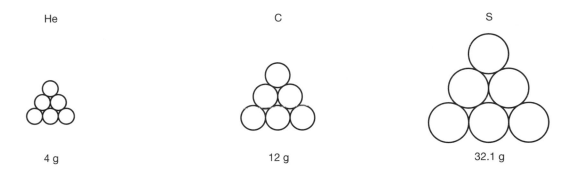

It is convenient for the purpose of this discussion to picture each of the above quantities as a pile of atoms, ignoring the normal structure of the elements.

Now consider **1 atom** of each of these elements, writing their masses in 'atomic mass units' (amu). (Note that, strictly speaking, atomic masses are **relative** atomic masses with no unit. However, the use of 'amu' may help the understanding of the point being made here.)

$$\begin{array}{ccc} \text{He} & \text{C} & \text{S} \\ 4\ \text{amu} & 12\ \text{amu} & 32.1\ \text{amu} \end{array}$$

Look at the pile of atoms representing 1 mol of He, and weighing 4 g. If 1 atom of He weighs 4 amu, a moment's thought tells us that the number of atoms in the pile is equal to the number of times that 4 amu goes into 4 g. That is:

$$\text{Number of atoms of He in 1 mol} = \frac{4\ \text{g}}{4\ \text{amu}} = \frac{1\ \text{g}}{1\ \text{amu}}$$

Applying the same thinking to the piles of carbon and sulphur atoms, each pile representing 1 mol of the element, we can say that:

$$\text{Number of atoms of C in 1 mole} = \frac{12\,\text{g}}{12\,\text{amu}} = \frac{1\,\text{g}}{1\,\text{amu}}$$

$$\text{Number of atoms of S in 1 mole} = \frac{32.1\,\text{g}}{32.1\,\text{amu}} = \frac{1\,\text{g}}{1\,\text{amu}}$$

In each of the above cases, **the number of atoms in a mole of the element is the same.** Its actual value has been calculated to be 6.02×10^{23} (to 3 significant figures) and is known as the Avogadro Constant, symbol L. Its formal unit is mol^{-1} which should be interpreted as '(particles) per mol'.

As stated above, we can now use the term 'mole' to mean an actual number of particles, with a numerical value of 6.02×10^{23}. However, as will be seen, great care must be taken to specify the particular type of particle that we are referring to. For example, the word 'dozen' refers to an actual number, but it is clear that 'a dozen hydrogen atoms' is different from 'a dozen hydrogen molecules'. (The former has an atomic mass of 12 and contains 12 H atoms; the latter has a molecular mass of 24 and contains 12 H_2 molecules, equal to a total of 24 H atoms.) Similarly, if we refer to a 'mole of hydrogen' we must be very careful to say whether we mean hydrogen atoms or molecules.

Consider the following examples carefully to ensure that this point is clear.

1 mol of C	contains	6.02×10^{23} **atoms** of C
1 mol of H_2	contains	6.02×10^{23} **molecules** of H_2
	contains	12.04×10^{23} **atoms** of H
1 mol of CO_2	contains	6.02×10^{23} **molecules** of CO_2
	contains	6.02×10^{23} **atoms** of C
	contains	12.04×10^{23} **atoms** of O
1 mol of $CaCl_2$	contains	6.02×10^{23} Ca^{2+} **ions**
	contains	12.04×10^{23} Cl^- **ions**
	contains	18.06×10^{23} **ions** in total
1 mol of Ne	contains	6.02×10^{23} Ne **atoms**
	contains	60.02×10^{23} **protons**
		(since each Ne atom contains 10 protons)

Note: In the above examples, the usual convention about writing numbers in Standard Form with a number between 1 and 10 multiplied by 10 to some power has, in some cases, been ignored to make comparison between the numbers easier. For example, 12.04×10^{23} is written rather than 1.204×10^{24}. In all future examples, however, the usual convention will be observed.

Worked Example 7.1

How many atoms are present in 0.3 mol of ammonia, NH_3?

We start with a statement connecting 1 mol of ammonia with the Avogadro Constant.

$$1 \text{ mol of } NH_3 = 6.02 \times 10^{23} \text{ molecules of } NH_3$$

As the question refers to **atoms,** this statement is rewritten:

$$1 \text{ mol of } NH_3 = 4 \times 6.02 \times 10^{23} \text{ atoms (of N and H)}$$
$$= 2.408 \times 10^{24} \text{ atoms}$$

(since 1 molecule of NH_3 contains 4 atoms (1 N and 3 H))

$$\text{So } 0.3 \text{ mol of } NH_3 = 0.3 \times 2.408 \times 10^{24} \text{ atoms}$$
$$= 7.224 \times 10^{23} \text{ atoms}$$
$$= 7.22 \times 10^{23} \text{ (to 3 significant figures)}$$

Worked Example 7.2

A sample of methane, CH_4, contains 1.204×10^{23} hydrogen atoms. How many moles of methane does the sample contain?

$$1 \text{ mol of } CH_4 = 6.02 \times 10^{23} \text{ molecules of } CH_4$$
$$\text{contains } 4 \times 6.02 \times 10^{23} \text{ atoms of H}$$
$$= 2.408 \times 10^{24} \text{ atoms of H}$$

This statement now connects 'mol of CH_4' with 'atoms of H'. In order to get our answer, in 'mol of CH_4', to come out on the right hand side, the statement is reversed:

2.408×10^{24} atoms of H are present in 1 mol of CH_4

1 atom of H is present in $\dfrac{1}{2.408 \times 10^{24}}$ mol of CH_4

1.204×10^{23} atoms of H are present in $\dfrac{1.204 \times 10^{23}}{2.408 \times 10^{24}}$ mol of CH_4

$$= 0.05 \text{ mol of } CH_4$$

PROBLEMS

The problems below are of the type illustrated by Worked Examples 7.1 and 7.2.

7.1 How many atoms are present in 1 mol of hydrogen gas, H_2?

7.2 How many hydrogen atoms are present in 1 mol of ammonia, NH_3?

7.3 How many molecules are present in 1.2 mol of glucose, $C_6H_{12}O_6$?

7.4 A sample of oxygen gas, O_2, contains 2.408×10^{21} atoms. How many moles of O_2 are present?

7.5 How many moles of methane, CH_4, would contain 9.03×10^{22} hydrogen atoms?

7.6 How many oxide ions are present in 0.2 mol of sodium oxide, Na_2O?

7.7 How many molecules are present in 0.2 mol of water, H_2O?

7.8 A sample of neon contains 3.612×10^{23} atoms. How many moles of neon does this represent?

7.9 How many ions are present in 0.5 mol of sodium oxide, Na_2O?

7.10 A beaker of dilute sulphuric acid, H_2SO_4, contains 4.816×10^{22} hydrogen ions. How many moles of sulphuric acid are present?

7.11 How many atoms of hydrogen are present in 0.02 mol of methanoic acid, HCOOH?

7.12 A sample of calcium chloride, $CaCl_2$, contains a total of 1.806×10^{20} ions. How many moles of calcium chloride are present?

7.13 How many ions are present in 2.25 mol of a solution of dilute sulphuric acid, H_2SO_4?

7.14 A sample of ethane, C_2H_6, contains 2.408×10^{24} atoms of carbon. How many moles of ethane are present?

7.15 How many ions are present in 0.4 mol of ammonium sulphate, $(NH_4)_2SO_4$?

7.16 A sample of aluminium nitrate, $Al(NO_3)_3$, contains 6.02×10^{24} nitrate ions. How many moles of the aluminium nitrate are present?

7.17 How many atoms are present in 2.4 mol of hydrogen sulphide, H_2S?

7.18 How many ions are present in 0.15 mol of magnesium fluoride, MgF_2?

7.19 A sample of ammonium sulphate, $(NH_4)_3PO_4$ contains 1.204×10^{23} ions. How many moles of the compound are present in the sample?

7.20 How many hydrogen atoms are present in 0.04 mol of propene, C_3H_6?

Worked Example 7.3

How many atoms are present in 1.215 g of magnesium?
The problem here involves **two** meanings of the mole, **mass** and **number**, so we state these meanings below, as applied to Mg.

$$1 \text{ mol of Mg} \qquad = 24.3 \text{ g}$$
$$1 \text{ mol of Mg} \qquad = 6.02 \times 10^{23} \text{ atoms of Mg}$$

Since these two statements both say '1 mole of Mg = ...' on the left hand side, the two right hand sides must also be the same. We can thus write:

$$24.3 \text{ g} \qquad = 6.02 \times 10^{23} \text{ atoms of Mg}$$

OR, expressing this the other way round,

$$6.02 \times 10^{23} \text{ atoms of Mg} = 24.3 \text{ g}$$

Since we want the final answer to come out as a number of atoms on the right hand side, we choose the first of the two forms:

$$24.3 \quad \text{g} \qquad = 6.02 \times 10^{23} \qquad \text{atoms of Mg}$$

$$1 \quad \text{g} \qquad = \frac{6.02 \times 10^{23}}{24.3} \qquad \text{atoms of Mg}$$

$$1.215 \text{ g} \qquad = \frac{1.215 \times 6.02 \times 10^{23}}{24.3} \quad \text{atoms of Mg}$$

$$= 3.01 \times 10^{22} \text{ atoms of Mg}$$

Worked Example 7.4

How many oxygen atoms are present in 0.22 g of carbon dioxide, CO_2?

As before, statements defining the two appropriate definitions of the mole are written.

$$1 \text{ mol of } CO_2 \qquad = 44 \text{ g}$$
$$1 \text{ mol of } CO_2 \qquad = 6.02 \times 10^{23} \text{ molecules of } CO_2$$

The second of these statements requires to be adjusted to suit the question which refers, not to **molecules of CO_2**, but to **atoms of O**. Since each molecule of CO_2 contains 2 atoms of O, we write:

$$1 \text{ mole of } CO_2 \qquad \text{contains } 1.204 \times 10^{24} \text{ atoms of O}$$

Since this statement and the first one above both refer to '1 mol of CO_2 ...' we can say:

$$44 \quad \text{g of } CO_2 \text{ contains } 1.204 \times 10^{24} \quad \text{atoms of O}$$

$$1 \quad \text{g of } CO_2 \text{ contains } \frac{1.204 \times 10^{24}}{44} \quad \text{atoms of O}$$

$$0.22 \text{ g of } CO_2 \text{ contains } \frac{0.22 \times 1.204 \times 10^{24}}{44} \quad \text{atoms of O}$$

$$= 6.02 \times 10^{21} \text{ atoms of O}$$

Worked Example 7.5

What mass of ammonia, NH_3 would contain 3.612×10^{22} atoms of hydrogen?

$$1 \text{ mol of } NH_3 = \quad 17 \text{ g}$$
$$1 \text{ mol of } NH_3 = \quad 6.02 \times 10^{23} \text{ molecules of } NH_3$$
$$\text{contains } 3 \times 6.02 \times 10^{23} \text{ atoms of H}$$
$$= \quad 1.806 \times 10^{24} \text{ atoms of H}$$

Connecting the first and last of these statements, with 'g of NH_3' on the right, we have:

$$1.806 \times 10^{24} \text{ atoms of H are contained in 17} \quad \text{g of } NH_3$$

$$1 \quad \text{atom of H is contained in } \frac{17}{1.806 \times 10^{24}} \quad \text{g of } NH_3$$

$$3.612 \times 10^{22} \text{ atoms of H are contained in } \frac{17 \times 3.612 \times 10^{22}}{1.806 \times 10^{24}} \text{ g of } NH_3$$

$$= 0.34 \text{ g of } NH_3$$

Worked Example 7.6

What is the mass, in grams, of 1 molecule of ammonia?

This problem is actually no different from the previous ones connecting the mass of a substance and a number of particles, although the fact that the number of particles referred to is 1 (molecule of ammonia) is sometimes confusing. We proceed exactly as before.

$$1 \text{ mol of } NH_3 = 17 \text{ g}$$
$$1 \text{ mol of } NH_3 = 6.02 \times 10^{23} \text{ molecules of } NH_3$$

We require our answer to come out as a mass, in grams on the right hand side, so we connect the above statements as below:

6.02×10^{23} molecules of NH_3 = 17 g

1   molecule of NH_3 = $\dfrac{17}{6.02 \times 10^{23}}$ g

= 2.82×10^{-23} g (to 3 significant figures)

PROBLEMS

These problems are of the type illustrated by Worked Examples 7.4–7.6 involving masses and numbers of particles.

7.21 How many atoms are present in 0.202 g of neon?

7.22 What is the mass of 1.505×10^{22} atoms of carbon?

7.23 How many atoms are present in 3.27 g of zinc?

7.24 What is the mass of 4.816×10^{21} atoms of chromium?

7.25 What is the mass, in grams, of 1 atom of gold?

7.26 How many atoms are present in 7.82 g of potassium?

7.27 What is the mass of 1.204×10^{21} atoms of magnesium?

7.28 How many atoms are present in 4 g of hydrogen, H_2?

7.29 What is the mass of 3.01×10^{21} molecules of nitrogen, N_2?

7.30 What is the mass, in grams, of 1 molecule of carbon dioxide, CO_2?

7.31 How many atoms are present in 2.64 g of carbon dioxide, CO_2?

7.32 A sample of hydrogen gas, H_2, contains 1.505×10^{21} atoms. What is the mass of the sample?

7.33 How many ions are present in 4.44 g of calcium chloride?

7.34 What is the mass of 1.204×10^{22} molecules of methane, CH_4?

7.35 How many *(a)* sodium ions and *(b)* sulphate ions are present in 11.368 g of sodium sulphate, Na_2SO_4?

7.36 What is the mass, in grams, of 1 molecule of ozone, O_3?

7.37 A sample of ethene, C_2H_4 weighs 2.24 g. How many carbon atoms will it contain?

7.38 A solution of iron(III) nitrate, $Fe(NO_3)_3$, is made by dissolving 12.09 g of the substance in water. How many nitrate ions are present in the solution?

7.39 How many atoms are present in 9 g of glucose, $C_6H_{12}O_6$?

7.40 A sample of aluminium oxide, Al_2O_3, contains 1.204×10^{21} oxide ions. What mass is the sample of aluminium oxide?

8

Gas Densities and Molar Volumes

The density of a substance is the mass of 1 unit of volume of the material. Several different units of density are used, such as $g\,cm^{-3}$ ('grams per cubic centimetre'), $g\,l^{-1}$ ('grams per litre'), $kg\,m^{-3}$ ('kilograms per cubic metre'), etc. In this chapter, only the unit of $g\,l^{-1}$ will be used.

The equation defining density is:

$$\text{Density} = \frac{\text{Mass of Sample}}{\text{Volume of Sample}}$$

The other two related equations are:

$$\text{Volume of Sample} = \frac{\text{Mass of Sample}}{\text{Density}}$$

$$\text{Mass of Sample} = \text{Volume of Sample} \times \text{Density}$$

The relationship between these three equations can be worked out by putting the quantities into the triangle shown below. (This can be the same method as that described in Chapter 1 for connecting concentration, moles and volume of solution.)

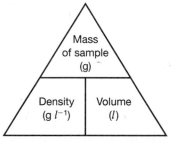

As can be seen, the positions of the three quantities in the equations are the same as they are in the triangle. So only one equation or the position of the quantities in the triangle needs to be remembered to be able to work out the other equations.

Now consider the four gases listed below, with their densities given in units of $g\,l^{-1}$. These figures apply at Standard Temperature and Pressure (stp), that is 0 °C and 1 atmosphere of pressure.

Gas	N_2	O_2	CO	He
Density ($g\,l^{-1}$)	1.25	1.43	1.25	0.179

We can calculate the volume occupied by 1 mol of a gas (the 'molar volume') as follows, taking N_2 as an example.

The mass of 1 mol of N_2 = 28 g

The equation to obtain the volume of 1 mol is:

$$\text{volume} = \frac{\text{mass}}{\text{density}}$$

$$\text{volume} = \frac{28}{1.25}$$

$$= 22.4\,l$$

This same calculation can be carried out using the 'direct variation' method used throughout the book, as follows.

From the density value, we can write:

1.25 g   occupies a volume of $1\,l$

1 g   occupies a volume of $\dfrac{1}{1.25}\,l$

28 g (1 mol)   occupies a volume of $\dfrac{28}{1.25}\,l$

$$= 22.4\,l$$

If the same calculation is carried out for all the gases listed, it is found that they all have approximately the same molar volume, around 22.4 l (at stp). Since the density figures given above only apply at 0 °C and 1 atmosphere of pressure, and since most industrial chemistry involves gases at other temperatures and pressures, the figure of 22.4 l is not a particularly useful one to remember. **What is important, however, is that 1 mol of *any* gas occupies very nearly the same volume when measured under the same conditions of temperature and pressure.**

Calculations involving density and molar volume data can be carried out using the 'triangle' equations or direct variation. In the former case, where a **molar** volume is involved, the triangle and related equations are as below:

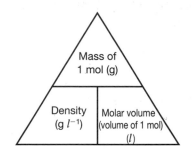

$$\text{Density} = \frac{\text{Mass of 1 Mol}}{\text{Molar Volume}}$$

$$\text{Mass of 1 mol} = \text{Density} \times \text{Molar Volume}$$

$$\text{Molar Volume} = \frac{\text{Mass of 1 mol}}{\text{Density}}$$

In the Worked Examples, both the methods described are shown; students are advised to work through both to find out which they find easier.

Worked Example 8.1

Under certain conditions of temperature and pressure, the density of methane, CH_4, is $1.06 \, g \, l^{-1}$. Calculate the molar volume of the gas under these conditions.

Both methods for tackling this problem will be shown; the first using an equation to connect density, molar volume and the mass of 1 mol, and the second using 'direct variation'.

Method 1: Using Equations
The appropriate form of the equation is selected and the data inserted.

$$\text{molar volume} = \frac{\text{mass of 1 mol}}{\text{density}}$$

$$= \frac{16}{1.06}$$

$$= \mathbf{15.1} \, l$$

(to 3 significant figures)

Method 2: Using 'Direct Variation'
The first step is to turn the density figure into an opening statement with 'litres' on the right hand side, since we want our answer to turn out to be a volume.

$$1.06 \text{ g} \qquad \text{is the mass of } 1 \quad l$$

$$1 \quad \text{g} \qquad \text{is the mass of } \frac{1}{1.06} \ l$$

$$16 \quad \text{g (1 mol) is the mass of } \frac{16}{1.06} \ l$$

$$= 15.1 \ l$$
(to 3 significant figures)

Worked Example 8.2

Under certain conditions, sulphur dioxide gas, SO_2, has a molar volume of 26 l. Calculate the density of the gas under these conditions.

Method 1: Using Equations

$$\text{density} = \frac{\text{mass of 1 mol}}{\text{molar volume}}$$

$$= \frac{64.1}{26}$$

$$= 2.47 \text{ g } l^{-1}$$
(to 3 significant figures)

Method 2: Using 'Direct Variation'

$$26 \ l \text{ is the volume of } 64.1 \text{ g of } SO_2 \text{ (1 mol)}$$

$$1 \ l \text{ is the volume of } \frac{64.1}{26} \text{ g of } SO_2$$

$$= 2.47 \text{ g}$$
(to 3 significant figures)

So the density of the gas is 2.47 g l^{-1}

Worked Example 8.3

A gas has a molar volume of 35 l and a density of 0.8 g l^{-1} under certain conditions. Calculate the molecular mass of the gas.

Method 1: Using Equations

$$\text{mass of 1 mol} = \text{density} \times \text{molar volume}$$

$$= 0.8 \times 35$$

$$= 28 \text{ g}$$

Since the mass of 1 mol of the gas is 28 g, its molecular mass is 28.

Method 2: Using 'Direct Variation'

$$1 \; l \qquad\qquad \text{has a mass of } 0.8 \qquad g$$
$$35 \; l \; (1 \; \text{mol}) \quad \text{has a mass of } 35 \times 0.8 \; g$$
$$= 28 \; g$$

Since the mass of 1 mol of the gas is 28 g, its molecular mass is 28.

PROBLEMS

The following problems are of the type illustrated by Worked Examples 8.1, 8.2 and 8.3.

8.1 At 0 °C and at atmospheric pressure, chlorine gas, Cl_2, has a density of $3.2 \, g \, l^{-1}$. Calculate the molar volume of chlorine under these conditions.

8.2 Under certain conditions, oxygen gas, O_2, has a molar volume of 25 l. Calculate the density of the gas under these conditions.

8.3 The industrial production of ammonia by the Haber Process requires nitrogen, N_2, and hydrogen, H_2 at a temperature of 450 °C and at 250 atmospheres. Under these conditions, the densities of nitrogen and hydrogen are, respectively, $118 \, g \, l^{-1}$ and $8.53 \, g \, l^{-1}$. Calculate the molar volume of each gas under these conditions.

8.4 At 0 °C and 1 atmosphere, a gas has a molar volume of 22.4 l and a density of $0.71 \, g \, l^{-1}$. Calculate the molecular mass of the gas.

8.5 Hydrogen sulphide, H_2S, has a molar volume of 65 l under certain conditions. Calculate the density of the gas under these conditions.

8.6 Under certain conditions, neon, Ne, has a density of $0.9 \, g \, l^{-1}$. Calculate the molar volume of the gas under these conditions.

8.7 A gas has a molar volume of 29.1 l and a density of $2.2 \, g \, l^{-1}$ under certain conditions of temperature and pressure. Calculate the molecular mass of the gas.

8.8 Ethyne, C_2H_2, has a molar volume of 50 l under certain conditions. What will the density of the gas be under these conditions?

8.9 A gaseous hydrocarbon has a molar volume of 17.3 l and a density of $1.735 \, g \, l^{-1}$ at a certain temperature and pressure. Calculate the molecular mass of the gas and hence identify it.

8.10 At 20 °C and atmospheric pressure, a diatomic gaseous element has a molar volume of 24.1 l and a density of $1.33 \, g \, l^{-1}$. Calculate the molecular mass of the gas and hence identify it.

Worked Example 8.4

A 25 cm^3 sample of a gas weighs 0.0221 g. If its molar volume under these conditions is 29.4 l, calculate its molecular mass.

Method 1: Using Equations
The first step is to calculate the density of the gas:

$$\text{density} = \frac{\text{mass}}{\text{volume}}$$

$$= \frac{0.0221}{0.025} \text{ (volume changed to } l \text{ from cm}^3\text{)}$$

$$= 0.884 \text{ g } l^{-1}$$

We then select the appropriate equation:

$$\text{mass of 1 mol} = \text{molar volume} \times \text{density}$$

$$= 29.4 \times 0.884$$

$$= 26.0 \text{ g}$$
$$\text{(to 3 significant figures)}$$

So the molecular mass of the gas is 26.0.

Method 2: Using 'Direct Variation'.
To obtain the mass of 1 mol of the gas, in grams, we need an opening statement with grams on the right hand side.

After converting 25 cm^3 into litres, we have:

$0.025 \ l$ is the volume of 0.0221 g of the gas

$1 \quad l$ is the volume of $\dfrac{0.0221}{0.025}$ g of the gas

$29.4 \quad l$ (1 mol) is the volume of $\dfrac{0.0221 \times 29.4}{0.025}$ g of the gas

$$= 26.0 \text{ g}$$
$$\text{(to 3 significant figures)}$$

So the molecular mass of the gas is 26.0.

Worked Example 8.5

The molar volume of carbon dioxide, CO_2, at stp, is 22.4 l. Calculate the mass of a 50 cm^3 sample of the gas under these conditions.

Method 1: Using Equations

$$\text{density} = \frac{\text{mass of 1 mol}}{\text{molar volume}}$$

$$= \frac{44}{22.4} \qquad g\,l^{-1}$$

$$= 1.964 \qquad g\,l^{-1} \text{ (rounded)}$$

We then put this calculated value for density, and the volume of the sample given, into the equation below.

$$\text{mass} = \text{density} \times \text{volume}$$

$$= 1.964 \times 0.05 \text{ (volume changed from cm}^3 \text{ to } l\text{)}$$

$$= 0.0982\,g \text{ (to 3 significant figures)}$$

Method 2: Using 'Direct Variation'

22.4 l is the volume of 44 g of CO_2 (1 mol)

1 l is the volume of $\dfrac{44}{22.4}$ g of CO_2

0.05 l is the volume of $\dfrac{44 \times 0.05}{22.4}$ g of CO_2

$$= 0.0982 \quad g \text{ (to 3 significant figures)}$$

Worked Example 8.6

The molar volume of nitrogen gas, N_2, is 25 l under certain conditions. What volume would 0.4 g of nitrogen gas occupy under the same conditions?

Method 1: Using Equations

$$\text{density} = \frac{\text{mass of 1 mol}}{\text{molar volume}}$$

$$= \frac{28}{25} \quad g\,l^{-1}$$

$$= 1.12\,g\,l^{-1}$$

This calculated value for density, and the mass of the sample given, are put into the equation below.

$$\text{volume} = \frac{\text{mass}}{\text{density}}$$

$$= \frac{0.4}{1.12} \, l$$

$$= 0.357 \, l \text{ (to 3 significant figures)}$$

Method 2: Using 'Direct Variation'

28 g (1 mol) is the mass of 25 l of N_2

1 g   is the mass of $\dfrac{25}{28} l$ of N_2

0.4 g   is the mass of $\dfrac{25 \times 0.4}{28} l$ of N_2

$$= 0.357 \, l \text{ (to 3 significant figures)}$$

PROBLEMS

These problems are of the type illustrated by Worked Examples 8.4, 8.5 and 8.6.

8.11 At 0 °C and 1 atmosphere of pressure, a 60 cm³ sample of carbon monoxide, CO, weighs 0.075 g. Calculate its molar volume under these conditions.

8.12 Ethyne, C_2H_2, has a molar volume of 22 l under certain conditions. Calculate the mass of 20 cm³ of ethyne under these conditions.

8.13 Helium, He, has a molar volume of 54 l under certain conditions. What volume would 0.24 g of helium occupy under the same conditions?

8.14 A 40 cm³ sample of hydrogen, H_2, weighs 0.0008 g under certain conditions. Calculate the molar volume of hydrogen under these conditions.

8.15 Under certain conditions, the molar volume of ammonia, NH_3, is 13.8 l. Calculate the mass of 150 l of this gas under these conditions.

8.16 Sulphur trioxide, SO_3, has a molar volume of 24 l under certain conditions. Calculate the volume which 0.15 g of the gas would occupy under these conditions.

8.17 A 10 cm³ sample of xenon, Xe, weighs 0.054 g at room temperature and pressure. Calculate its molar volume under these conditions.

8.18 Under certain conditions, sulphur dioxide has a molar volume of 58.2 *l*. Calculate the mass of 1000 *l* of the gas under these conditions.

8.19 Butane, C_4H_{10}, has a molar volume of 32 *l* under certain conditions of temperature and pressure. Calculate the volume which 0.045 g of the gas would occupy under the same conditions.

8.20 A 100 cm³ flask is filled with carbon dioxide, CO_2, at room temperature and pressure. The mass of the gas in the flask is found to be 0.181 g. Calculate the molar volume of the gas under these conditions.

Worked Example 8.7

Carbon dioxide has a molar volume of 32 *l* under certain conditions of temperature and pressure. 50 cm³ of a sample of an unknown monatomic gas under the same conditions weighed 0.0625 g. Calculate the atomic mass of the gas, and hence identify it.

We are told the molar volume of carbon dioxide and asked about another gas under the same conditions. The assumption that we need to make is that **both gases have the same molar volume under the same conditions.** So we can say that the molar volume of the unknown gas is also 32 *l*.

We proceed as below.

Method 1: Using Equations
We have the information to calculate the density of the gas.

$$\text{density} = \frac{\text{mass}}{\text{volume}}$$

$$= \frac{0.0625}{0.05} \text{ (volume changed to } l \text{ from cm}^3\text{)}$$

$$= 1.25 \text{ g } l^{-1}$$

We then use this, and the molar volume in the appropriate equation:

$$\text{mass of 1 mole} = \text{molar volume} \times \text{density}$$

$$= 32 \times 1.25$$

$$= 40 \text{ g}$$

The only monatomic gases are the Noble Gases; **argon,** with a relative atomic mass of 40, is the only one which fits the data.

Method 2: Using 'Direct Variation'

We start with the same assumption used in the first method, namely that the molar volume of our unknown gas is the same as that of carbon dioxide under the same conditions, that is, $32\,l$.

We want to obtain the atomic mass of the gas; another way of saying this is that we want to know the mass of 1 mol of the gas (e.g. if the atomic mass is 15, then the mass of 1 mol will be 15 g). So we start with a statement which will make our answer come out as a mass on the right hand side of the problem.

After firstly converting our $50\,cm^3$ into litres, we have:

$$0.05\ l \qquad \text{has a mass of} \qquad 0.0625 \quad g$$

$$1 \quad l \qquad \text{has a mass of} \qquad \frac{0.0625}{0.05} \quad g$$

$$32 \quad l\,(1\ \text{mol}) \ \ \text{has a mass of} \quad \frac{32 \times 0.0625}{0.05} \quad g$$

$$= 40\,g$$

Our conclusion is the same as for Method 1. The only monatomic gas (Noble Gas) with a relative atomic mass of 40 is **argon**.

PROBLEMS

The following problems are of the type illustrated by Worked Example 8.7.

8.21 Under certain conditions, nitrogen, N_2, has a density of $0.68\,g\,l^{-1}$.
   (a) Calculate the molar volume of nitrogen under these conditions.
   (b) Calculate the densities of oxygen and carbon dioxide under the same conditions.

8.22 $50\,cm^3$ of ethane, C_2H_6, weighs 0.0625 g; calculate the molar volume of the gas under these conditions. Under the same conditions, $50\,cm^3$ of an unknown gas is found to weigh 0.0792 g. Calculate the molecular mass of the unknown gas.

8.23 Under certain conditions, a $200\,cm^3$ sample of oxygen, O_2, weighs 0.358 g. A $400\,cm^3$ sample of a gas, X, weighs 0.354 g under the same conditions. Calculate the molecular mass of X.

8.24 A $200\,cm^3$ sample of hydrogen sulphide, H_2S, weighs 0.268 g. Under the same conditions of temperature and pressure, $40\,cm^3$ of an unknown gas weighed 0.05 g. Calculate the molecular mass of the unknown gas.

8.25 A $20\,cm^3$ sample of krypton, Kr, weighs 0.0672 g. What mass would a $50\,cm^3$ sample of argon, Ar, have under identical conditions of temperature and pressure?

9

Calculations from Equations – 4: Involving Gas Volumes

In this chapter, we consider calculations from equations where the only chemical species we are concerned with are gases under the same conditions of temperature and pressure, and where the quantities of gas are expressed as volumes. Because a mole of any gas occupies the same volume under the same conditions, we do not have to calculate actual numbers of moles in this type of problem; we simply use the volumes of gases since these values are proportional to the numbers of moles.

Worked Example 9.1

What volume of carbon dioxide would be produced if 20 cm^3 of ethane is burned in an excess of oxygen? (All volumes measured under the same conditions of temperature and pressure.)

We start the problem as in previous calculations from equations:

Step 1: Balanced Equation

$$C_2H_6(g) \ + \ 3\tfrac{1}{2}O_2(g) \longrightarrow 2CO_2(g) \ + \ 3H_2O(l)$$

Step 2: Mole Statement

1 mol of C$_2$H$_6$ will produce 2 mol of CO$_2$

Step 3: Volume Statement

Since 1 mol of **any gas** occupies the same volume under the same conditions of temperature and pressure, we can rewrite our mole statement in terms of 'volumes' as below, where the 'volumes' can be in any unit of volume provided that the same unit is used consistently throughout the problem:

1 volume of C$_2$H$_6$ will produce 2 volumes of CO$_2$

Step 4: Finishing Off

From the volume statement above, we can say that since:

1 volume of C$_2$H$_6$ will produce 2 volumes of CO$_2$,

20 cm^3 of C$_2$H$_6$ will produce **40 cm^3 of CO$_2$**

Worked Example 9.2

What would be the volume and composition of the resulting gas mixture if $100 \, cm^3$ of butane was exploded in $800 \, cm^3$ of oxygen? (All volumes are measured at 20 °C and at 1 atmosphere of pressure.)

Step 1: Balanced Equation

$$C_4H_{10}(g) \; + \; 6\tfrac{1}{2}O_2(g) \longrightarrow 4CO_2(g) \; + \; 5H_2O(l)$$

Steps 2 and 3: Mole and Volume Statements

The problem refers to the volumes of **gases**; at 20 °C and 1 atmosphere pressure (roughly room conditions), water is a liquid and all the other species in the equation are gases, as indicated by the use of the 'state subscripts' (*l* and g) in the equation. We therefore ignore the quantity of water involved since the problem only asks about the resulting **gas mixture** and write statements as below.

1 mol   of C_4H_{10} will react with $6\tfrac{1}{2}$ mol   of O_2 to form 4 mol   of CO_2

1 volume of C_4H_{10} will react with $6\tfrac{1}{2}$ volumes of O_2 to form 4 volumes of CO_2

Step 4: Finishing Off

Consideration of the volume statement and the available volumes of butane and oxygen tells us that there is more oxygen present than will react (it is in excess). All $100 \, cm^3$ of butane will react as follows:

1 volume of C_4H_{10} will react with $6\tfrac{1}{2}$ volumes of O_2 to form 4 volumes of CO_2,

$100 \, cm^3$ of C_4H_{10} will react with $650 \, cm^3$ of O_2 to form $400 \, cm^3$ of CO_2.

Thus the final mixture will contain **$400 \, cm^3$ of CO_2**. Since we started with $800 \, cm^3$ of oxygen but only $650 \, cm^3$ of it has reacted, there will also be **$150 \, cm^3$ of O_2** unreacted in the final gas mixture.

PROBLEMS

These problems involve gases measured under the same conditions of temperature and pressure and should be worked out by the method described in Worked Examples 9.1 and 9.2.

9.1 $CH_4(g) \; + \; 2O_2(g) \longrightarrow CO_2(g) \; + \; 2H_2O(l)$
        $100 \, cm^3$ of methane is exploded with $300 \, cm^3$ of oxygen.
    (*a*)   Which gas is in excess?
    (*b*)   What is the volume and composition of the resulting gas mixture?

9.2 $$CO(g) + \tfrac{1}{2}O_2(g) \longrightarrow CO_2(g)$$

50 cm^3 of carbon monoxide is burned with 20 cm^3 of oxygen.

LIBRARY

Sighthill Campus

Opening Hours
(Term-time)

Library
Monday- Friday
8:30 - 16:30

PCs and Macs must be booked at the library desk and will shutdown 15 minutes before the library closes

Borrowing
(ID card required)

You may borrow:
6 books for three weeks each
4 magazines for one week each
3 DVDs for one week each

Fines
Books/journals: 10p per item per day
DVDs: 50p per item per day

Contact
Library
Edinburgh College - Sighthill Campus
Bankhead Avenue
Edinburgh
EH11 4DE

0131 297 9996
library@edinburghcollege.ac.uk

composition of the resulting gas mixture?

$$(g) \longrightarrow 3CO_2(g) + 4H_2O(l)$$

140 l of oxygen. Calculate the volume and gas mixture.

$$(g) \longrightarrow 3CO_2(g) + 3H_2O(l)$$

ded with 5000 l of oxygen. Calculate the volume ing gas mixture.

$$(g) \longrightarrow 2CO_2(g) + 3H_2O(g)$$

completely in 2 l of oxygen. Calculate the volume ing gas mixture if all volume measurements were °C and a pressure of 1 atmosphere; i.e. under the gas state (steam).

$$(g) \longrightarrow N_2(g) + 2H_2O(g)$$

urned in 400 l of oxygen to form nitrogen gas and and composition of the resulting gas mixture be if at 300 °C and at the same pressure?

$$(g) \longrightarrow 4CO_2(g) + 5H_2O(l)$$

000 l of oxygen. What is the volume and gas mixture? (All volume measurements taken at a tmosphere pressure.)

$$(g) \longrightarrow 6CO_2(g) + 13H_2(g)$$

m hexane by catalytic reaction with steam. Under f hexane is reacted with an excess of steam. completion, calculate the volume of (a) carbon duced.

$$3) \longrightarrow CCl_4(g) + S_2Cl_2(g)$$

be made industrially by the above process in which th chlorine gas. Under certain conditions, 5×10^5 l of d completely in the presence of 2.5×10^6 l of chlorine. lume of the resulting gas mixture?

$$(g) \longrightarrow 6CO(g) + 7H_2(g)$$

sed to carbon monoxide and oxygen gases. Under l of hydrogen gas was obtained from this reaction. ne, measured under the same conditions, must have

10

Calculations from Equations – 5: Involving Gas Densities and Molar Volumes of Gases

In this chapter we consider calculations from equations involving the molar volumes and densities of gases. The work of Chapter 8 should be revised before continuing.

Worked Example 10.1

8.8 g of propane is completely burned in an excess of oxygen. What volume of carbon dioxide would be produced if its density under these conditions is $2\,\mathrm{g}\,l^{-1}$.

Step 1: Balanced Equation
$$C_3H_8 + 5O_2 \longrightarrow 3CO_2 + 4H_2O$$

Step 2: Mole Statement
$$1 \text{ mol of } C_3H_8 \text{ produces 3 mol of } CO_2$$

Step 3: Calculation of 'Known' Moles
The known substance is propane, C_3H_8. 1 mol of C_3H_8 = 44 g.

$$44 \quad \mathrm{g} = 1 \quad \text{mol of } C_3H_8$$

$$1 \quad \mathrm{g} = \frac{1}{44} \text{ mol of } C_3H_8$$

$$8.8 \ \mathrm{g} = \frac{8.8}{44} \text{ mol of } C_3H_8$$

$$= 0.2 \ \text{mol of } C_3H_8$$

Step 4: Calculation of 'Unknown' Moles
$$1 \quad \text{mol of } C_3H_8 \text{ produces} \quad 3 \text{ mol of } CO_2$$
$$0.2 \ \text{mol of } C_3H_8 \text{ produces } 0.2 \times 3 \text{ mol of } CO_2$$
$$= 0.6 \text{ mol of } CO_2$$

Step 5: Finishing Off
We have calculated that:

$$8.8 \text{ g of } C_3H_8 \text{ would produce } 0.6 \text{ mol of } CO_2$$

We have to obtain the **volume** of CO_2 and are given that the density of the gas is $2\,\mathrm{g}\,l^{-1}$ under the conditions of the reaction.

We convert the 0.6 mol of CO_2 into grams:

$$
\begin{aligned}
1 \quad \text{mol of } CO_2 &= \quad 44 \quad g \\
0.6 \quad \text{mol of } CO_2 &= 0.6 \times 44 \ g \\
&= 26.4 \quad g
\end{aligned}
$$

Using the density figure, we have:

2 g of CO_2 has a volume of $\quad 1 \quad l$

1 g of CO_2 has a volume of $\quad \dfrac{1}{2} \quad l$

26.4 g of CO_2 has a volume of $\dfrac{26.4}{2} \ l$

$$= 13.2 \quad l$$

PROBLEMS

● Problems 10.1 to 10.10 are of the type illustrated by Worked Example 10.1, involving either molar volume or density.

10.1 $$CH_4(g) + 2O_2(g) \longrightarrow CO_2(g) + 2H_2O(g)$$
7.2 g of methane is burned completely in oxygen according to the above equation. What volume of carbon dioxide would be produced if the molar volume of this gas is 200 l under these conditions?

10.2 $$Na_2CO_3(s) + 2HCl(aq) \longrightarrow 2NaCl(aq) + CO_2(g) + H_2O(l)$$
5.3 g of sodium carbonate is reacted with an excess of dilute hydrochloric acid. What volume of carbon dioxide will be given off if the density of the gas is 2 g l^{-1}?

10.3 $$4NH_3(g) + 5O_2(g) \longrightarrow 4NO(g) + 6H_2O(l)$$
400 l of oxygen is used up in the above reaction, measured under conditions in which its density is 1.6 g l^{-1}. What mass of ammonia must have reacted?

10.4 $$3F_2(g) + 3H_2O(g) \longrightarrow 6HF(g) + O_3(g)$$
1824 g of fluorine is reacted completely with an excess of steam. What volume of ozone gas, O_3, would be produced under conditions where its density is 2.4 g l^{-1}?

10.5 $$Zn(s) + 2HCl(aq) \longrightarrow ZnCl_2(aq) + H_2(g)$$
Under certain conditions, the molar volume of hydrogen gas is 24 l. What mass of zinc would be required to react with an excess of hydrochloric acid to produce 4 l of hydrogen under these conditions?

10.6 $H_2SO_4(aq) + CaCO_3(s) \longrightarrow CaSO_4(aq) + CO_2(g) + H_2O(l)$
What volume of carbon dioxide would be given off when 20 g of calcium carbonate is completely reacted with an excess of sulphuric acid? The molar volume of carbon dioxide is 22 l under the reaction conditions.

10.7 $N_2H_4(g) + 2F_2(g) \longrightarrow N_2(g) + 4HF(g)$
What volume of hydrogen fluoride would be produced by the complete reaction of 6.4 kg of hydrazine, N_2H_4, with an excess of fluorine? The molar volume of hydrogen fluoride is 50 l under the conditions of measurement.

10.8 $Fe_2O_3(s) + 3CO(g) \longrightarrow 2Fe(s) + 3CO_2(g)$
Under conditions in which the molar volume of carbon dioxide is 100 l, 1500 l of the gas is obtained by the reduction of iron(III) oxide by carbon monoxide. What mass of iron(III) oxide must have been reduced?

10.9 $3Cu(s) + 8HNO_3(aq) \longrightarrow 3Cu(NO_3)_2(aq) + 4H_2O(l) + 2NO(g)$
15.24 g of copper is completely reacted with an excess of nitric acid according to the above equation. Assuming that this is the only reaction taking place, calculate the volume of nitrogen monoxide which would be evolved if its density under the reaction conditions is 1.6 g l^{-1}.

10.10 $C_6H_{14}(g) + 6H_2O(g) \longrightarrow 6CO(g) + 13H_2(g)$
$1.08 \times 10^6 \, l$ of hexane, C_6H_{14}, is reacted with an excess of steam according to the above equation. If the molar volume of the hexane under these conditions is 120 l, calculate the mass of hydrogen which would be produced.

Calculations from Equations – 6: Involving Percentages

In many chemical processes, the reactions do not go 'to completion' with all the reactants being used up to give 100% 'yield'. One reason for this is that many reactions have a 'back reaction' as well as a forward reaction; under certain circumstances, an equilibrium can be set up where the forward and back reactions take place at the same rate.

Another factor is that many chemical processes involve starting materials which are **impure**; that is, only a certain percentage of the original mass is the actual reactant, the rest of the mass being impurities which do not react to make the desired product.

In this chapter we will consider reactions where there is not a 100% yield. In some problems, the percentage yield will be given, and a mass of product will be asked for. In others, the actual mass reacting or being produced will be given and the percentage yield of product (compared to the theoretical maximum yield) will be asked for.

The methods used in these calculations can be seen in the following Worked Examples. They follow the methods described in Chapter 2, with a percentage calculation included.

Worked Example 11.1

Ethene reacts with chlorine to form dichloroethane. If 17.5 g of ethene is introduced to the reaction container with an excess of chlorine, but only 80% of the ethene reacts, calculate the mass of dichloroethane produced.

Step 1: Balanced Equation

$$C_2H_4 + Cl_2 \longrightarrow C_2H_4Cl_2$$

Step 2: Mole Statement
1 mol of C_2H_4 reacts to form 1 mol of $C_2H_4Cl_2$

This statement assumes that the reaction goes to completion; that is, that **all** the C_2H_4 will react.

Step 3: Calculation of 'Known' Moles

The 'known' substance is C_2H_4 and the mass given is 17.5 g. But we are told that only 80% of it reacts.

So the mass **actually** reacting is 80% of 17.5 g = $0.8 \times 17.5 = \mathbf{14\,g}$

We then carry out the rest of the calculation exactly as in Chapter 2.

1 mol of C_2H_4 = 28 g.

$$28 \text{ g} = 1 \quad \text{mol}$$

$$1 \text{ g} = \frac{1}{28} \quad \text{mol}$$

$$14 \text{ g} = 14 \times \frac{1}{28} \text{ mol}$$

$$= 0.5 \quad \mathbf{mol}$$

Step 4: Calculation of 'Unknown' Moles

The 'unknown' is $C_2H_4Cl_2$

From Step 1, we know that:

$$1 \text{ mol of } C_2H_4 \text{ reacts to form 1 mol of } C_2H_4Cl_2$$

So

$$0.5 \text{ mol of } C_2H_4 \text{ reacts to form } \mathbf{0.5 \text{ mol of } C_2H_4Cl_2}$$

Step 5: Finishing Off

$$1 \text{ mol } \text{ of } C_2H_4Cl_2 = 99 \text{ g}$$
$$0.5 \text{ mol of } C_2H_4Cl_2 = 0.5 \times 99 \text{ g}$$
$$= \mathbf{49.5\,g}$$

So the mass of dichloroethane produced is 49.5 g.

PROBLEMS

These problems of the type illustrated by Worked Example 11.1 in which a stated percentage of a product reacts.

In Problems 11.5–11.10, 'industrial' quantities in kg are used, and often expressed in 'Standard Form', e.g. 3×10^4 kg. If the use of these quantities is unfamiliar, it is recommended that reference should be made to Chapter 2, Worked Example 2.2.

11.1
$$Fe_2O_3 + 3CO \longrightarrow 2Fe + 3CO_2$$
Iron(III) oxide can be reduced to iron by carbon monoxide. 19.95 g of iron(III) is treated with carbon monoxide, but only 80% by mass reacts. What mass of iron will be produced?

11.2
$$CaCO_3 \longrightarrow CaO + CO_2$$
Calcium carbonate can be decomposed to calcium oxide by heat. A form of rock which is 60% calcium carbonate by mass is crushed and heated until all the calcium carbonate has decomposed. If 40 g of rock is so treated, what mass of calcium oxide could theoretically be obtained?

11.3
$$CH_4 + H_2O \longrightarrow 3H_2 + CO$$
Methane can be reacted with steam to form hydrogen and carbon monoxide although, under the reaction conditions, only 30% methane, by mass, reacts. Under these conditions, if 4 g of methane is introduced to the reaction chamber, what mass of hydrogen would be produced?

11.4
$$2PbS + 3O_2 \longrightarrow 2PbO + 2SO_2$$
The mineral galena, lead(II) sulphide, can be converted to lead(II) oxide by reaction with oxygen. If 47.86 g of galena is so treated, but there is only a 75% yield, what mass of lead(II) oxide would be produced?

11.5
$$TiCl_4 + 2Mg \longrightarrow Ti + 2MgCl_2$$
In the conversion of titanium(IV) chloride to titanium, using an excess of magnesium, a 60% yield of product is obtained under certain conditions. If 37.98 g of titanium(IV) chloride is so treated, what mass of titanium would be obtained?

11.6
$$N_2 + 3H_2 \longrightarrow 2NH_3$$
In the manufacture of ammonia by the Haber Process, under certain conditions, there is only a 70% yield of product. If 8.40×10^4 kg of nitrogen is reacted with an excess of hydrogen under these conditions, calculate the mass of ammonia produced.

11.7
$$C_6H_{12}O_6 \longrightarrow 2C_2H_5OH + 2CO_2$$
The fermentation of glucose to ethanol and carbon dioxide produces a 75% yield of ethanol under certain conditions. If 72 kg of glucose is fermented under these conditions, what mass of ethanol would be produced?

11.8
$$2SO_2 + O_2 \longrightarrow 2SO_3$$
In the manufacture of sulphur trioxide from sulphur dioxide, as a preliminary stage in the manufacture of sulphuric acid, 7.68×10^4 kg of sulphur dioxide is introduced to the reaction chamber in the presence of an excess of oxygen over a period of time. If the reaction goes to 75% completion, calculate the mass of sulphur trioxide produced in this time.

11.9
$$2ZnS + 3O_2 \longrightarrow 2ZnO + 2SO_2$$
Zinc sulphide can be converted by the above reaction to form zinc oxide, from which zinc metal can then be extracted. If 1.95×10^4 kg of ore which is known to contain 40% zinc oxide by mass is so treated, what mass of zinc oxide would be obtained? (Assume that all the available zinc sulphide reacts according to the above equation.)

11.10
$$CH_3COOH + CH_3OH \longrightarrow CH_3COOCH_3 + H_2O$$
In the above esterification of ethanoic acid with methanol to form methyl ethanoate and water, 750 kg of ethanoic acid, CH_3COOH, is heated with an excess of methanol in the presence of a sulphuric acid catalyst. If there is a 72% yield of product, calculate the mass of methyl ethanoate which this represents.

Worked Example 11.2

3 g of ethanoic acid, CH_3COOH, is reacted with an excess of ethanol in the presence of a concentrated sulphuric acid catalyst. This results in 2.64 g of the ester, ethyl ethanoate, being produced. Express this as a percentage yield of product.

In this problem, we firstly assume that **all** the ethanoic acid reacts and calculate the mass of ethyl ethanoate which would be produced. The mass of product **actually** made is then expressed as a percentage of this theoretical mass.

Step 1: Balanced Equation
$$CH_3COOH + C_2H_5OH \longrightarrow CH_3COOC_2H_5 + H_2O$$

Step 2: Mole Statement
$$1 \text{ mol of } CH_3COOH \text{ reacts to form } 1 \text{ mol of } CH_3COOC_2H_5$$

This statement assumes that the reaction goes to completion; that is, that **all** the CH_3COOH will react.

Step 3: Calculation of 'Known' Moles
The 'known' substance is CH_3COOH. 1 mol = 60 g.

$$60 \text{ g} = 1 \quad \text{mol}$$

$$1 \text{ g} = \frac{1}{60} \quad \text{mol}$$

$$3 \text{ g} = \frac{3 \times 1}{60} \text{ mol}$$

$$= 0.05 \quad \text{mol}$$

Step 4: Calculation of 'Unknown' Moles

The 'unknown' is $CH_3COOC_2H_5$

From Step 1, we know that:

$$1 \text{ mol of } CH_3COOH \text{ reacts to form 1 mol of } CH_3COOC_2H_5$$

So

$$0.05 \text{ mol of } CH_3COOH \text{ reacts to form } 0.05 \text{ mol of } CH_3COOC_2H_5$$

So, if the reaction had gone to completion, **0.05 mol of $CH_3COOC_2H_5$** would have been formed.

Step 5: Finishing Off

$$
\begin{aligned}
1 \quad \text{mol of } CH_3COOC_2H_5 &= 88\,g \\
0.05 \text{ mol of } CH_3COOC_2H_5 &= 0.05 \times 88 \ g \\
&= 4.4\,g
\end{aligned}
$$

This is the mass of $CH_3COOC_2H_5$ which would have been formed, **had all the original mass of ethanoic acid reacted.**

We are told in the problem that the **actual** mass of $CH_3COOC_2H_5$ which is produced is 2.64 g.

$$
\begin{aligned}
\text{The } \textbf{percentage yield} &= \frac{\text{Actual mass obtained}}{\text{Maximum theoretical mass}} \times 100\% \\
&= \frac{2.64}{4.4} \times 100\% \\
&= 60\%
\end{aligned}
$$

So the yield of product in this reaction is 60%.

PROBLEMS

These problems are of a type similar to Worked Example 11.2, in which percentage yield of a product or percentage purity of a reactant require to be calculated. In problems 11.16–11.20, masses are expressed in 'industrial' quantities in kg and often expressed in 'Standard Form', eg. 3×10^4 kg.

11.11
$$CO + 2H_2 \longrightarrow CH_3OH$$
Carbon monoxide and hydrogen gas can be reacted in the presence of a catalyst to form methanol, CH_3OH. Under particular reaction conditions, 22.4 kg of

carbon monoxide is mixed with an excess of hydrogen. After a time, 19.2 kg of methanol has been produced.

(a) Calculate the mass of methanol which would have been formed *had all the carbon monoxide reacted*.

(b) Express the mass of methanol *actually produced* as a percentage (the 'percentage yield') of the theoretical mass calculated in part (a).

11.12
$$C_2H_4 + HI \longrightarrow C_2H_5I$$
Ethene reacts with hydrogen iodide to form iodoethane, C_2H_5I. 7 g of ethene is mixed with an excess of hydrogen iodide in a reaction container for a period of time, after which 15.59 g of iodoethane was obtained. Calculate the percentage yield of product, by mass, for this process.

11.13
$$2FeCl_2 + Cl_2 \longrightarrow 2FeCl_3$$
In the reaction represented by the above equation, 5.072 g of iron(II) chloride yields 4.869 g of iron(III) chloride under certain conditions. Calculate the percentage yield of product.

11.14
$$C_6H_6 + HNO_3 \longrightarrow C_6H_5NO_2 + H_2O$$
18.72 g of benzene, C_6H_6, enters a reaction chamber with an excess of nitric acid. After a time, 22.14 g of nitrobenzene, $C_6H_5NO_2$, is obtained. Calculate the percentage yield of product.

11.15
$$Na_2CO_3 + H_2SO_4 \longrightarrow Na_2SO_4 + CO_2 + H_2O$$
A 5.3 g sample of impure sodium carbonate is reacted with an excess of sulphuric acid, resulting in 1.76 g of carbon dioxide being given off.

(a) Calculate the mass of pure sodium carbonate which must have been present in the sample.

(b) Express this as a percentage of the mass of the original, impure, sample. This is the 'percentage purity' of original sample. (Assume that the impurity does not react with the acid to give off carbon dioxide.)

11.16
$$2Cu_2S + 3O_2 \longrightarrow 2Cu_2O + 2SO_2$$
3.182×10^3 kg of copper(I) sulphide is heated with an excess of oxygen, producing 1.716×10^3 kg of copper(I) oxide. Express this latter mass as a percentage yield of product.

11.17
$$4NH_3 + 5O_2 \longrightarrow 4NO + 6H_2O$$
Ammonia can be oxidised in the presence of oxygen and a catalyst to produce nitrogen monoxide. In a certain period of time 1.36×10^3 kg of ammonia enters the reaction chamber and 1.68×10^3 kg of nitrogen monoxide is produced. Calculate the percentage yield of nitrogen monoxide over this time.

11.18
$$2SO_2 + O_2 \longrightarrow 2SO_3$$
The above equation represents the conversion of sulphur dioxide to sulphur trioxide. Over a period of time, 3.205×10^3 kg of sulphur dioxide is introduced

to the reaction chamber in the presence of an excess of oxygen. 2.403×10^3 kg of sulphur trioxide is obtained during the process; express the mass of product as a percentage yield.

11.19 $$Fe + 2HCl \longrightarrow FeCl_2 + H_2$$

A 4.65 kg batch of scrap metal, consisting mainly of iron, is analysed by being treated with an excess of hydrochloric acid, causing all the iron to be converted to iron(II) chloride solution. After evaporation of the solution and the removal of all other substances, 9.51 kg of pure, solid, iron(II) chloride was obtained.

(*a*) Calculate the mass of pure iron in the batch of scrap.

(*b*) Express the mass of pure iron in the scrap metal as a percentage of the mass of the original batch.

11.20 $$Fe_2O_3 + 3CO \longrightarrow 2Fe + 3CO_2$$

6.65×10^4 kg of an iron ore which is impure iron(III) oxide is reacted with an excess of carbon monoxide, producing 2.79×10^4 kg of iron.

(*a*) Calculate the mass of pure iron(III) oxide in the ore. (Assume that all the iron(III) oxide is reduced to iron and that the impurities do not take part in the above reaction.)

(*b*) Calculate the percentage, by mass, of iron(III) oxide in the ore.

Hess's Law

Hess's Law of thermochemistry states (in simplified form) that the enthalpy change (ΔH) for a reaction depends only on the enthalpies of the reactants and the products and not on how the reaction is carried out or on how many steps are involved in the process.

An implication of this law is that if we can rearrange and combine chemical equations in such a way that we obtain a different equation, we can obtain the ΔH for the new reaction from the ΔH values of the original reactions. This idea seems complicated, but the following Worked Examples will clarify it.

Worked Example 12.1

Calculate the ΔH for the reaction below:

$$C\ (s)\ +\ 2H_2\ (g) \longrightarrow CH_4\ (g)$$

Use the following data:

❶	$C\ (s)\ +\ O_2\ (g) \longrightarrow CO_2\ (g)$	$\Delta H = -394\ kJ\ mol^{-1}$
❷	$H_2\ (g)\ +\ \frac{1}{2}O_2\ (g) \longrightarrow H_2O\ (l)$	$\Delta H = -286\ kJ\ mol^{-1}$
❸	$CH_4\ (g)\ +\ 2O_2\ (g) \longrightarrow CO_2\ (g) + 2H_2O\ (l)$	$\Delta H = -891\ kJ\ mol^{-1}$

The basic technique is to rearrange the equations which we are told to use and to put them into a form which, when added together, will give us the required equation.

The equation we are trying to get to is the one below:

$$C\ (s)\ +\ 2H_2\ (g) \longrightarrow CH_4\ (g)$$

The first substance on the left hand side (LHS) of this equation is C (s) so our first step is to find an equation in the ones we are told to use which contains C (s). The equation labelled ❶ has C (s) on the LHS, so we write it just as it is with its ΔH value alongside.

❶	$C\ (s)\ +\ O_2\ (g) \longrightarrow CO_2\ (g)$	$\Delta H = -394\ kJ\ mol^{-1}$

The next substance that we want to help give us the required equation is $2H_2$ (g) on the LHS. Equation ❷ has H_2 (g) on the LHS, so this equation and its ΔH value are doubled. The equation is now represented as $2 \times$ ❷.

$$2 \times ❷ \quad 2H_2 \text{ (g)} + O_2 \text{ (g)} \longrightarrow 2H_2O \text{ (l)} \qquad\qquad \Delta H = -572 \text{ kJ mol}^{-1}$$

The last substance we require is CH_4 (g) on the right hand side (RHS). CH_4 (g) is present in equation ❸, but on the LHS, so we reverse this equation and change the sign of its ΔH value and write it as below as '$-$❸'.

$$-❸ \quad CO_2 \text{ (g)} + 2H_2O \text{ (l)} \longrightarrow CH_4 \text{ (g)} + 2O_2 \text{ (g)} \qquad \Delta H = +891 \text{ kJ mol}^{-1}$$

Collecting these three equations together, we can then cancel out species common to both sides* and add them, and their ΔH values, together as shown below.

❶	$C \text{ (s)} + O_2 \text{ (g)} \longrightarrow CO_2 \text{ (g)}$	$\Delta H = -394 \text{ kJ mol}^{-1}$
$2 \times$ ❷	$2H_2 \text{ (g)} + O_2 \text{ (g)} \longrightarrow 2H_2O \text{ (l)}$	$\Delta H = -572 \text{ kJ mol}^{-1}$
$-$❸	$CO_2 \text{ (g)} + 2H_2O \text{ (l)} \longrightarrow CH_4 \text{ (g)} + 2O_2 \text{ (g)}$	$\Delta H = +891 \text{ kJ mol}^{-1}$
	$C \text{ (s)} + 2H_2 \text{ (g)} \longrightarrow CH_4 \text{ (g)}$	$\Delta H = -75 \text{ kJ mol}^{-1}$

*The cancelling took place as follows:

- Two O_2 (g), one from the LHS of the first equation and one from the LHS of the second equation, cancel with the $2O_2$ (g) on the RHS of the third equation.
- $2H_2O$ (l) appears on the RHS of the second equation and on the LHS of the third. They are cancelled.
- CO_2 (g) appears on the RHS of the first equation and on the LHS of the third equation. They are cancelled.

The equation that we obtain after the adding up and cancelling of the rearranged equations is exactly the one that we require.

So the calculated ΔH value of -75 kJ mol^{-1} is the required enthalpy change.

In Worked Example 10.1, all the equations were provided. However, often, energy terms are expressed as recognised terms such as Enthalpy of Combustion and Enthalpy of Formation.

Enthalpy of Combustion

The Enthalpy of Combustion is defined as the amount of energy given out when 1 mol of a substance is burned completely in oxygen. All these values refer to the energy values of the substances being measured at their usual, room temperature, states. Very often the Enthalpy of Combustion refers to a compound containing carbon and hydrogen, or carbon hydrogen **and** oxygen.

When carbon burns completely in oxygen it forms carbon dioxide. When hydrogen burns completely in oxygen it forms water.

The Enthalpies of Combustion of carbon and hydrogen are represented by the equations and ΔH values shown below.

$$C\ (s)\ +\ O_2\ (g)\ \longrightarrow\ CO_2\ (g) \qquad\qquad \Delta H = -394\ \text{kJ mol}^{-1}$$

$$H_2\ (g)\ +\ \tfrac{1}{2}O_2\ (g)\ \longrightarrow\ H_2O\ (l) \qquad\qquad \Delta H = -286\ \text{kJ mol}^{-1}$$

Note: In the equation above showing the combustion of hydrogen, the amount of O_2 required is shown as $\tfrac{1}{2}O_2$ (g). This means **half a mole of oxygen gas – not** half an oxygen molecule. Multiples of oxygen molecules involving $\tfrac{1}{2}$, eg $1\tfrac{1}{2}O_2$ (g), $3\tfrac{1}{2}O_2$ (g) etc. will occur frequently in this chapter.

When a compound containing carbon and hydrogen, or carbon, hydrogen and oxygen, burns completely, carbon dioxide **and** water are formed.

For example, the Enthalpy of Combustion of methane, CH4, is represented by the equation and ΔH value shown below.

$$CH_4\ (g)\ +\ 2O_2\ (g)\ \longrightarrow\ CO_2\ (g)\ +\ 2H_2O(l) \qquad \Delta H = -891\ \text{kJ mol}^{-1}$$

Enthalpy of Formation
The Enthalpy of Formation of a compound is the enthalpy change when 1 mol of the substance is formed from its elements. As with the Enthalpy of Combustion, all substances involved are in their usual, room temperature, states.

For example, the Enthalpy of Formation of Ethanol, C_2H_5OH, refers to the equation and ΔH value below.

$$2C\ (s)\ +\ 3H_2\ (g)\ +\ \tfrac{1}{2}O_2\ (g)\ \longrightarrow\ C_2H_5OH\ (l) \qquad \Delta H = -278\ \text{kJ mol}^{-1}$$

This definition of Enthalpy of Formation does not require to be learned for Higher Chemistry; where it is referred to in this book, the required equations will be provided.

Note that the SQA Data Book, p 9 contains a table with Standard Enthalpies of Formation and Combustion of selected substances. This table is also in Appendix 2 on p 142 of this book. **Care should be taken not to take information from the wrong column when referring to this table.**

Worked Example 12.2

The formation of methanol from its elements is represented by the equation below:

$$C\ (s)\ +\ 2H_2\ (g)\ +\ \tfrac{1}{2}O_2\ (g)\ \longrightarrow\ CH_3OH\ (l)$$

Calculate the ΔH for this reaction using the Enthalpies of Combustion of carbon, hydrogen and methanol, obtained from Appendix 2 on page 142.

The three combustion equations and their ΔH values, obtained from Appendix 2 are written and labelled as ❶, ❷ and ❸ for ease of reference.

❶ $C\ (s)$ $+\ O_2\ (g)\ \longrightarrow\ CO_2\ (g)$ $\Delta H = -394\ \text{kJ mol}^{-1}$

❷ $H_2\ (g)$ $+\ \frac{1}{2}O_2\ (g)\ \longrightarrow\ H_2O\ (l)$ $\Delta H = -286\ \text{kJ mol}^{-1}$

❸ $CH_3OH\ (l)$ $+\ 1\frac{1}{2}O_2\ (g)\ \longrightarrow\ CO_2\ (g)\ +\ 2H_2O\ (l)$ $\Delta H = -727\ \text{kJ mol}^{-1}$

They are then rewritten below, multiplying and/or reversing them to suit the form of the required equation for the formation of methanol. The ΔH values, multiplied and/or with changed signs where necessary, are written alongside.

❶ $C\ (s)$ $+\ O_2\ (g)$ $\longrightarrow\ CO_2\ (g)$ $\Delta H = -394\ \text{kJ mol}^{-1}$

$2 \times$ ❷ $2H_2\ (g)\ +\ O_2\ (g)$ $\longrightarrow\ 2H_2O\ (l)$ $\Delta H = -572\ \text{kJ mol}^{-1}$

$-$ ❸ $CO_2\ (g)\ +\ 2H_2O\ (l)\ \longrightarrow\ CH_3OH\ (l) + 1\frac{1}{2}O_2\ (g)$ $\Delta H = +727\ \text{kJ mol}^{-1}$

The rearranged equations and their ΔH values can then be added up, after cancelling of species common to both sides ($O_2\ (g)$, $CO_2(g)$ and $H_2O\ (l)$), to give:

$$C\ (s)\ +\ 2H_2\ (g)\ +\ \tfrac{1}{2}O_2\ (g)\ \longrightarrow\ CH_3OH\ (l) \qquad \Delta H = -239\ \text{kJ mol}^{-1}$$

This is the required equation and therefore the value for the Enthalpy of Formation of methanol is $-239\ \text{kJ mol}^{-1}$.

Note that, in writing these equations, there was no choice in how to get $C\ (s)$ on the LHS, $2H_2(g)$ on the LHS and $CH_3OH\ (l)$ on the RHS. However, if we had attempted to get $\frac{1}{2}O_2\ (g)$ on the LHS directly by selecting one equation, we would have had difficulty, since $O_2\ (g)$ appears in all three equations and we would not necessarily be sure which one(s) to use. The problem in this case was solved by ignoring it! If all three equations given can be written or rewritten to get the $C\ (s)$, $H_2O\ (l)$ and $CH_3OH\ (l)$ in the right amounts and on the correct sides of the equation, then the $O_2\ (g)$, being the only remaining species, **must** work out correctly.

This is a bit like doing a jigsaw puzzle, where you don't know where to put one particular piece; if you can put all the other pieces in their correct places, the final one **must** fit into the last remaining space.

PROBLEMS

- In problems 12.1 to 12.10 all the required equations and data are given in the problem.
- In problems 12.11 to 12.20, some equations will need to be worked out and some data will need to be obtained from p. 9 of the SQA Data Book or Appendix 2 of this book. The method, however, is exactly the same as in the Worked Examples and earlier problems in this chapter.

12.1 The equation for the combustion of methane, CH_4, is given below:

$$CH_4 \text{ (g)} + 2O_2 \text{ (g)} \longrightarrow CO_2 \text{ (g)} + 2H_2O \text{ (l)}$$

Calculate the Enthalpy of Combustion of methane, using the data below.

C (s)	$+ O_2$ (g)	$\longrightarrow CO_2$ (g)	$\Delta H = -394 \text{ kJ mol}^{-1}$
H_2 (g)	$+ \frac{1}{2}O_2$ (g)	$\longrightarrow H_2O$ (l)	$\Delta H = -286 \text{ kJ mol}^{-1}$
C (s)	$+ 2H_2$ (g)	$\longrightarrow CH_4$ (g)	$\Delta H = -75 \text{ kJ mol}^{-1}$

12.2 The equation representing the combustion of ethane, C_2H_6, is given below:

$$C_2H_6 \text{ (g)} + 3\frac{1}{2}O_2 \text{ (g)} \longrightarrow 2CO_2 \text{ (g)} + 3H_2O \text{ (l)}$$

Calculate the Enthalpy of Combustion of ethane, C_2H_6, using the Enthalpy of Formation of ethane and the Enthalpies of Combustion of carbon and hydrogen. The equations and ΔH values for these processes are given below:

2C (s)	$+ 3H_2$ (g)	$\longrightarrow C_2H_6$ (g)	$\Delta H = -85 \text{ kJ mol}^{-1}$
C (s)	$+ O_2$ (g)	$\longrightarrow CO_2$ (g)	$\Delta H = -394 \text{ kJ mol}^{-1}$
H_2 (g)	$+ \frac{1}{2}O_2$ (g)	$\longrightarrow H_2O$ (l)	$\Delta H = -286 \text{ kJ mol}^{-1}$

12.3 The Enthalpy of Formation of ethane, C_2H_6, is represented by the following equation:

$$2C \text{ (s)} + 3H_2 \text{ (g)} \longrightarrow C_2H_6 \text{ (g)}$$

Calculate the Enthalpy of Formation of ethane using the Enthalpies of Combustion of hydrogen, carbon and ethane, represented by the equations and ΔH values below:

C (s)	$+ O_2$ (g)	$\longrightarrow CO_2$ (g)	$\Delta H = -394 \text{ kJ mol}^{-1}$
H_2 (g)	$+ \frac{1}{2}O_2$ (g)	$\longrightarrow H_2O$ (l)	$\Delta H = -286 \text{ kJ mol}^{-1}$
C_2H_6 (g)	$+ 3\frac{1}{2}O_2$ (g)	$\longrightarrow 2CO_2$ (g) $+ 3H_2O$ (l)	$\Delta H = -1560 \text{ kJ mol}^{-1}$

12.4 The Enthalpy of Combustion of propane, C_3H_8, is the ΔH for the following reaction:

$$C_3H_8 \text{ (g)} + 5O_2 \text{ (g)} \longrightarrow 3CO_2 \text{ (g)} + 4H_2O \text{ (l)}$$

Calculate the Enthalpy of Combustion of propane using the Enthalpy of Formation of propane and the Enthalpies of Combustion of carbon and hydrogen. The equations and ΔH values for these processes are given below:

3C (s)	$+ 4H_2$ (g)	$\longrightarrow C_3H_8$ (g)	$\Delta H = -104 \text{ kJ mol}^{-1}$
C (s)	$+ O_2$ (g)	$\longrightarrow CO_2$ (g)	$\Delta H = -394 \text{ kJ mol}^{-1}$
H_2 (g)	$+ \frac{1}{2}O_2$ (g)	$\longrightarrow H_2O$ (l)	$\Delta H = -286 \text{ kJ mol}^{-1}$

12.5 The formation of butane, C_4H_{10}, is represented by the equation below:

$$4C \text{ (s)} + 5H_2 \text{ (g)} \longrightarrow C_4H_{10} \text{ (g)}$$

Calculate the Enthalpy of Formation of butane using the Enthalpies of Combustion of butane, hydrogen and carbon represented by the equations and ΔH values below:

$$C_4H_{10} \text{ (g)} + 6\tfrac{1}{2}O_2 \text{ (g)} \longrightarrow 4CO_2 \text{ (g)} + 5H_2O \text{ (l)} \qquad \Delta H = -2877 \text{ kJ mol}^{-1}$$
$$H_2 \text{ (g)} + \tfrac{1}{2}O_2 \text{ (g)} \longrightarrow H_2O \text{ (l)} \qquad \Delta H = -286 \text{ kJ mol}^{-1}$$
$$C \text{ (s)} + O_2 \text{ (g)} \longrightarrow CO_2 \text{ (g)} \qquad \Delta H = -394 \text{ kJ mol}^{-1}$$

12.6 The equation for the combustion of ethanol, C_2H_5OH, is:

$$C_2H_5OH \text{ (l)} + 3O_2 \text{ (g)} \longrightarrow 2CO_2 \text{ (g)} + 3H_2O \text{ (l)}$$

Calculate the Enthalpy of Combustion of ethanol using the Enthalpies of Combustion of carbon and hydrogen, and the Enthalpy of Formation of ethanol represented by the equations and ΔH values below:

$$C \text{ (s)} + O_2 \text{ (g)} \longrightarrow CO_2 \text{ (g)} \qquad \Delta H = -394 \text{ kJ mol}^{-1}$$
$$H_2 \text{ (g)} + \tfrac{1}{2}O_2 \text{ (g)} \longrightarrow H_2O \text{ (l)} \qquad \Delta H = -286 \text{ kJ mol}^{-1}$$
$$2C \text{ (s)} + 3H_2 \text{ (g)} + \tfrac{1}{2}O_2 \text{ (g)} \longrightarrow C_2H_5OH \text{ (l)} \qquad \Delta H = -278 \text{ kJ mol}^{-1}$$

12.7 The formation of ethanoic acid, CH_3COOH, is represented by the equation below:

$$2C \text{ (s)} + 2H_2 \text{ (g)} + O_2 \text{ (g)} \longrightarrow CH_3COOH \text{ (l)}$$

Calculate the Enthalpy of Formation of ethanoic acid using the Enthalpies of Combustion of carbon, hydrogen and ethanoic acid represented by the equations and ΔH values below:

$$C \text{ (s)} + O_2 \text{ (g)} \longrightarrow CO_2 \text{ (g)} \qquad \Delta H = -394 \text{ kJ mol}^{-1}$$
$$H_2 \text{ (g)} + \tfrac{1}{2}O_2 \text{ (g)} \longrightarrow H_2O \text{ (l)} \qquad \Delta H = -286 \text{ kJ mol}^{-1}$$
$$CH_3COOH \text{ (l)} + 2O_2 \text{ (g)} \longrightarrow 2CO_2 \text{ (g)} + 2H_2O \text{ (l)} \qquad \Delta H = -876 \text{ kJ mol}^{-1}$$

12.8 The formation of propan-1-ol, C_3H_7OH, is represented by the equation below:

$$3C \text{ (s)} + 4H_2 \text{ (g)} + \tfrac{1}{2}O_2 \text{ (g)} \longrightarrow C_3H_7OH \text{ (l)}$$

Calculate the Enthalpy of Formation of propan-1-ol using the Enthalpies of Combustion of carbon, hydrogen and propan-1-ol represented by the equations and ΔH values below:

$$C \text{ (s)} + O_2 \text{ (g)} \longrightarrow CO_2 \text{ (g)} \qquad \Delta H = -394 \text{ kJ mol}^{-1}$$
$$H_2 \text{ (g)} + \tfrac{1}{2}O_2 \text{ (g)} \longrightarrow H_2O \text{ (l)} \qquad \Delta H = -286 \text{ kJ mol}^{-1}$$
$$C_3H_7OH \text{ (l)} + 4\tfrac{1}{2}O_2 \text{ (g)} \longrightarrow 3CO_2 \text{ (g)} + 4H_2O \text{ (l)} \qquad \Delta H = -2020 \text{ kJ mol}^{-1}$$

12.9 The combustion of benzene, C_6H_6, is represented by the equation below.

$$C_6H_6 \ (l) \ + \ 7\tfrac{1}{2}O_2 \ (g) \longrightarrow 6CO_2 \ (g) \ + \ 3H_2O \ (l)$$

Calculate the Enthalpy of Combustion of benzene using the Enthalpies of Combustion of carbon and hydrogen and the Enthalpy of Formation of benzene represented by the equations and ΔH values below:

$C \ (s) \quad + \ O_2 \ (g) \longrightarrow CO_2 \ (g)$ $\qquad\qquad\qquad$ $\Delta H = -394 \ \text{kJ mol}^{-1}$

$H_2 \ (g) \ + \ \tfrac{1}{2}O_2 \ (g) \longrightarrow H_2O \ (l)$ $\qquad\qquad$ $\Delta H = -286 \ \text{kJ mol}^{-1}$

$6C \ (s) \ + \ 3H_2 \ (g) \longrightarrow C_6H_6 \ (l)$ $\qquad\qquad$ $\Delta H = +49 \ \text{kJ mol}^{-1}$

12.10 The formation of ethyne, C_2H_2, is represented by the equation below

$$2C \ (s) \ + \ H_2 \ (g) \longrightarrow C_2H_2 \ (g)$$

Calculate the Enthalpy of Formation of ethyne using the Enthalpies of Combustion of carbon, hydrogen and ethyne represented by the equations and ΔH values below:

$C \ (s) \quad + \ O_2 \ (g) \longrightarrow CO_2 \ (g)$ $\qquad\qquad\qquad$ $\Delta H = -394 \ \text{kJ mol}^{-1}$

$H_2 \ (g) \ + \ \tfrac{1}{2}O_2 \ (g) \longrightarrow H_2O \ (l)$ $\qquad\qquad$ $\Delta H = -286 \ \text{kJ mol}^{-1}$

$C_2H_2 \ (g) \ + \ 2\tfrac{1}{2}O_2 \ (g) \longrightarrow 2CO_2 \ (g) + H_2O \ (l)$ $\quad$ $\Delta H = -1300 \ \text{kJ mol}^{-1}$

12.11 The formation of methanoic acid, HCOOH, is represented by the following equation.

$$C \ (s) \ + \ H_2 \ (g) \ + \ O_2 \ (g) \longrightarrow HCOOH \ (l)$$

Calculate the ΔH for the above reaction using the Enthalpies of Combustion of carbon, hydrogen and methanoic acid.

12.12 The reaction of ethanol with oxygen to form ethanoic acid and water is represented by the equation below:

$$C_2H_5OH \ (l) \ + \ O_2 \ (g) \longrightarrow CH_3COOH \ (l) \ + \ H_2O \ (l)$$

Use the Enthalpies of Combustion of ethanol and ethanoic acid to calculate the ΔH for the above reaction.

12.13 The equation below represents the hydrogenation of ethene to ethane:

$$C_2H_4 \ (g) \ + \ H_2 \ (g) \longrightarrow C_2H_6 \ (g)$$

Use the Enthalpies of Combustion of ethene, hydrogen and ethane to calculate the ΔH for the above reaction.

12.14 The fermentation of glucose, $C_6H_{12}O_6$, to ethanol and carbon dioxide can be represented by the equation below:

$$C_6H_{12}O_6 \ (s) \longrightarrow 2C_2H_5OH \ (l) \ + \ 2CO_2 \ (g)$$

Calculate the ΔH for this reaction using the Enthalpies of Combustion of ethanol and glucose. (The Enthalpy of Combustion of glucose is $-2813 \ \text{kJ mol}^{-1}$.)

12.15 Use the Enthalpies of Combustion of ethyne, C_2H_2, ethane, C_2H_6, and hydrogen to calculate the ΔH for the complete hydrogenation of ethyne given by the equation below:

$$C_2H_2 \text{ (g)} + 2H_2 \text{ (g)} \longrightarrow C_2H_6 \text{ (g)}$$

12.16 The Enthalpy of Formation of diborane, B_2H_6, is the ΔH for the following reaction:

$$2B \text{ (s)} + 3H_2 \text{ (g)} \longrightarrow B_2H_6 \text{ (g)}$$

Calculate the Enthalpy of Formation of diborane using the Enthalpy of Combustion of hydrogen and the equations and ΔH values for the Enthalpy of Formation of boron oxide and the Enthalpy of Combustion of diborane noted below:

$2B \text{ (s)} \quad + 1\frac{1}{2}O_2 \text{ (g)} \longrightarrow B_2O_3 \text{ (s)}$	$\Delta H = -612 \text{ kJ mol}^{-1}$
$B_2H_6 \text{ (g)} + 3O_2 \text{ (g)} \longrightarrow B_2O_3 \text{ (s)} + 3H_2O \text{ (l)}$	$\Delta H = -1058 \text{ kJ mol}^{-1}$

12.17 Use the Enthalpies of Combustion of methane, hydrogen and ethyne, C_2H_2, to obtain the ΔH for the reaction represented by the equation below.

$$2CH_4 \text{ (g)} \longrightarrow C_2H_2 \text{ (g)} + 3H_2 \text{ (g)}$$

12.18

$Na \text{ (s)} + \frac{1}{2}Cl_2 \text{ (g)} \longrightarrow Na^+ \text{ (g)} + Cl^- \text{ (g)}$	$\Delta H = +365 \text{ kJ mol}^{-1}$
$Na \text{ (s)} + \frac{1}{2}Cl_2 \text{ (g)} \longrightarrow Na^+Cl^- \text{ (s)}$	$\Delta H = -411 \text{ kJ mol}^{-1}$

Use the above information to calculate the ΔH for the reaction below:

$$Na^+Cl^- \text{ (s)} \longrightarrow Na^+ \text{ (g)} + Cl^- \text{ (g)}$$

12.19

$Cu \text{ (s)} \longrightarrow Cu^{2+} \text{ (aq)} + 2e^-$	$\Delta H = +795 \text{ kJ mol}^{-1}$
$Cu \text{ (s)} \longrightarrow Cu^+ \text{ (aq)} + e^-$	$\Delta H = +602 \text{ kJ mol}^{-1}$

Use the above information to calculate the ΔH for the reaction below:

$$2Cu^+ \text{ (aq)} \longrightarrow Cu^{2+} \text{ (aq)} + Cu \text{ (s)}$$

12.20 The Enthalpies of Formation of methylhydrazine, CH_3NHNH_2, and dinitrogen tetroxide, N_2O_4, are represented by the equations and ΔH values below:

$C \text{ (s)} \quad + 3H_2 \text{ (g)} + N_2 \text{ (g)} \longrightarrow CH_3NHNH_2 \text{ (l)}$	$\Delta H = +53 \text{ kJ mol}^{-1}$
$N_2 \text{ (g)} + 2O_2 \text{ (g)} \longrightarrow N_2O_4 \text{ (g)}$	$\Delta H = -20 \text{ kJ mol}^{-1}$

Use this information, and the Enthalpies of Combustion of carbon and hydrogen, to obtain the ΔH for the equation below:

$$4CH_3NHNH_2 \text{ (l)} + 5N_2O_4 \text{ (g)} \longrightarrow 4CO_2 \text{ (g)} + 12H_2O \text{ (l)} + 9N_2 \text{ (g)}$$

pH Calculations

In earlier work at Standard Grade or Intermediate 2, the pH scale was introduced as a numerical scale from 0 to 14 which was used to describe how acid or alkaline a solution was.

- acid solutions had a pH of below 7
- alkali solutions had a pH of above 7
- neutral solutions had a pH of **exactly** 7

The pH scale is based on the mathematical relationship below:

$$pH = -\log[H^+(aq)]$$

In this equation, log is short for the mathematical function called the logarithm. The squared brackets around the H^+ refer to 'the concentration of H^+ in mol l^{-1}'. The meaning of the log function will become clear from considering a few examples.

$[H^+(aq)]$ (mol l^{-1})	$[H^+(aq)]$ (expressed as a power of 10) (mol l^{-1})	$\log[H^+(aq)]$	$pH = -\log[H^+(aq)]$
0.1	10^{-1}	-1	1
0.01	10^{-2}	-2	2
0.001	10^{-3}	-3	3
etc.			

As can be seen the log is simply the power of 10 to which a number has been raised (pH stands for the 'power of hydrogen'). The point of the pH scale may now be seen; it takes a sometimes clumsy decimal fraction and converts it into a number which is easier to handle. It is easier to say that a solution has a pH of 3 than to say that its hydrogen ion concentration is 0.001 mol l^{-1}. This is especially true when dealing with concentrations which are **non-whole number** powers of 10.

pH values of solutions with concentrations which can be expressed as whole number powers of 10 can be calculated by working out the power of 10 and taking its negative value. In the Worked Examples following, this is described as Method 1.

An alternative method (Method 2) is to use the log function on a scientific calculator. In the Worked Examples following, it is suggested that both the 'mental arithmetic' method (Method 1) **and** using a calculator (Method 2), should be followed to find the method which is more comfortable. Only limited guidance about entering information into the calculator is given here as the precise method will depend on the calculator used.

Worked Examples 13.1 and 13.2 and Problems 13.1 to 13.10 involve converting a concentration of hydrogen ions to a pH. Worked Examples 13.3 and 13.4 and Problems 13.11 to 13.20 involve the exact opposite process.

Worked Example 13.1

What is the pH of a solution whose hydrogen ion concentration is 10^{-4} mol l^{-1}?

Method 1 (Mental arithmetic)
Since the concentration is already expressed as a power of 10, the pH is simply the negative of this power, that is the negative of -4, which is 4.

So the pH is **4**.

Method 2 (Using the log function on a calculator)
The pH is the $-\log[H^+(aq)]$, so $-\log 10^{-4}$ is entered into the calculator. The display should read 4.

So the pH is **4**.

Worked Example 13.2

What is the pH of a solution whose hydrogen ion concentration is 0.001 mol l^{-1}?

Method 1 (Mental arithmetic)
This concentration firstly has to be expressed as a power of 10. The easiest way to do this, without using a calculator is to remember that 0.1 is 10^{-1} and simply count how many places the decimal point has been moved to go from 0.1 to 0.001. The point moved **two** places, so 0.001 is **two powers of 10 smaller** than 10^{-1}; the number is therefore 10^{-3}.
Since the pH is the negative of the power, the pH is the negative of -3, ie 3.

So the pH is **3**.

Method 2 (Using the log function on a calculator)
The pH is the $-\log[H^+(aq)]$, so $-\log 0.00001$ is entered into the calculator. The display should read 5.

So the pH is 5.

PROBLEMS

Problems 13.1 to 13.10 are of the type shown by Worked Examples 13.1 and 13.2, where the pH of a solution has to be calculated from the hydrogen ion concentration.

Calculate the pH of solutions where the concentration of H^+ is as stated below:

13.1 $[H^+(aq)] = 0.001 \text{ mol } l^{-1}$

13.2 $[H^+(aq)] = 10^{-6} \text{ mol } l^{-1}$

13.3 $[H^+(aq)] = 0.00001 \text{ mol } l^{-1}$

13.4 $[H^+(aq)] = 10^{-13} \text{ mol } l^{-1}$

13.5 $[H^+(aq)] = 0.0001 \text{ mol } l^{-1}$

13.6 $[H^+(aq)] = 10^{-8} \text{ mol } l^{-1}$

13.7 $[H^+(aq)] = 10^{-11} \text{ mol } l^{-1}$

13.8 $[H^+(aq)] = 10^{-4} \text{ mol } l^{-1}$

13.9 $[H^+(aq)] = 10^{-7} \text{ mol } l^{-1}$

13.10 $[H^+(aq)] = 10^{-14} \text{ mol } l^{-1}$

Worked Example 13.3

A solution has a pH of 9. What is its hydrogen ion concentration?

Method 1 (Mental arithmetic)
The pH is the **negative** of the power of 10 of the concentration. If the pH is 9, the power of 10 of the concentration is therefore the **negative of 9**, ie -9.

So the concentration of hydrogen ion is $10^{-9} \text{ mol } l^{-1}$.

Method 2 (Using the log function on a calculator)
The definition of pH is:

$$pH = -\log[H^+(aq)].$$

In earlier examples, to get the pH we took the log of the $[H^+(aq)]$ and then took the negative of that value, ie changed its sign.

To get the pH from the value of $[H^+(aq)]$ we do the **opposite**.

Firstly we take the negative of the pH to give us $\log[H^+(aq)]$.

The pH is 9, so the $\log[H^+(aq)]$ is -9.

Secondly, we 'unlog' this value of -9 by using the **inverse** or **second function** of log on the calculator.

This gives the value of 10^{-9}.

So the concentration of hydrogen ion is 10^{-9} mol l^{-1}.

PROBLEMS

Problems 13.11 to 13.20 are of the type shown by Worked Example 13.3 in which concentrations of hydrogen ions have to be calculated from pH values.

For each of the pH values below, calculate the concentration of hydrogen ions in the solution.

13.11 pH 5

13.12 pH 3

13.13 pH 7

13.14 pH 13

13.15 pH 2

13.16 pH 0

13.17 pH 4

13.18 pH 10

13.19 pH 12

13.20 pH 1

ADDITIONAL THEORY 1; The relationship between [H⁺(aq)] and [OH⁻(aq)] ions.

A feature of all aqueous solutions, whether acid, alkaline or neutral, is that they contain H^+ and OH^- ions.

- Acid solutions contain more H^+ than OH^-.
- Alkaline solutions contain more OH^- than H^+.
- Neutral solutions contain equal concentrations of H^+ than OH^-.

At 25 °C, the mathematical relationship between the concentrations of H^+ and OH^- in solution is:

$$[H^+(aq)][OH^-(aq)] = 10^{-14}\,mol^2\,l^{-2}$$

NOTE: The unit above is simply the unit of concentration $(mol\,l^{-1})$ **squared** since two concentration values, those of H^+ than OH^-, are being multiplied.

This relationship means that, if we know the concentration of H^+, we can calculate the concentration of OH^-, and *vice versa*.

This can be seen in the following examples.

Worked Example 13.4

What is the hydrogen ion concentration of an aqueous solution in which the hydroxide ion concentration is $0.001\,mol\,l^{-1}$ $(10^{-3}\,mol\,l^{-1})$?

The equation connecting the two ion concentrations is written.

$$[H^+(aq)][OH^-(aq)] = 10^{-14}\,mol^2\,l^{-2}$$

The known concentration, that of hydroxide, is inserted.

$$[H^+(aq)] \times 10^{-3} = 10^{-14}$$

Students familiar with multiplying numbers raised to powers will see immediately that

$$[H^+(aq)] = 10^{-11}\,mol\,l^{-1}$$

since $10^{-11} \times 10^{-3} = 10^{-14}$.

Students less confident about this can solve the equation for $[H^+(aq)]$ by writing.

$$[H^+(aq)] = \frac{10^{-14}}{10^{-3}}$$

and using a calculator to get

$$[H^+(aq)] = 10^{-11}\,mol\,l^{-1}$$

Worked Example 13.5

What is the pH of a solution with a hydroxide ion concentration of 10^{-5} mol l^{-1}?

In this problem we firstly have to calculate the concentration of hydrogen ions and then use that value to obtain the pH.

In the first part of the problem, we proceed as in the previous Worked Example:

$$[H^+(aq)][OH^-(aq)] = 10^{-14} \text{ mol}^2 \, l^{-2}$$

So $\qquad [H^+(aq)] \times 10^{-5} \quad = 10^{-14} \text{ mol}^2 \, l^{-2}$

So $\qquad [H^+(aq)] \qquad = \dfrac{10^{-14}}{10^{-5}} \text{ mol} \, l^{-1}$

So $\qquad [H^+(aq)] \qquad = 10^{-9} \text{ mol} \, l^{-1}$

The second part of the problem is exactly as in Worked Examples 13.1 and 13.2 (page 99) and Problems 13.1–13.10.

Using Method 1 the pH is seen as being the **negative of the power of 10** of the hydrogen ion concentration, which, in this case, is the **negative of −9**, ie **9**.

Using Method 2 since pH $= -\log[H^+(aq)]$, $-\log 10^{-9}$ is entered into the calculator, giving the value **9**.

So, using either method, the pH is 9.

Worked Example 13.6

The pH of a solution is 11. Calculate the concentration of hydroxide ions in the solution.

In this problem, the first step is to obtain the hydrogen ion concentration from the pH as in previous Worked Examples and Problems.

In this example, the pH is 11, so the $[H^+(aq)]$ is 10^{-11} mol l^{-1}

Note: If this calculation is still unclear, go back to Worked Example 13.3 and Problems 13.11–13.20.

We then proceed as before:

$$[H^+(aq)][OH^-(aq)] = 10^{-14} \ \text{mol}^2 \, l^{-2}$$

So $\qquad 10^{-11} \times [OH^-(aq)] = 10^{-14} \ \text{mol}^2 \, l^{-2}$

So $\qquad [OH^-(aq)] \qquad = \dfrac{10^{-14}}{10^{-11}} \ \text{mol} \, l^{-1}$

So $\qquad [OH^-(aq)] \qquad = 10^{-3} \ \text{mol} \, l^{-1}$

So the hydroxide ion concentration is $10^{-3} \, \text{mol} \, l^{-1}$.

Worked Example 13.7

What is (a) the hydrogen ion concentration and (b) the hydroxide ion concentration in a neutral solution, such as pure water?

The pH of a neutral solution is, of course, 7. So we can immediately work out that the hydrogen ion concentration is $10^{-7} \, \text{mol} \, l^{-1}$.

As before, we fit this value into the equation to obtain $[OH^-(aq)]$:

$$[H^+(aq)][OH^-(aq)] = 10^{-14} \ \text{mol}^2 \, l^{-2}$$

So $\qquad 10^{-7} \times [OH^-(aq)] = 10^{-14} \ \text{mol}^2 \, l^{-2}$

So $\qquad [OH^-(aq)] \qquad = \dfrac{10^{-14}}{10^{-7}} \ \text{mol} \, l^{-1}$

So $\qquad [OH^-(aq)] \qquad = 10^{-7} \ \text{mol} \, l^{-1}$

This Worked Example was included, not because it is any different from the previous one but because it illustrates what the neutral pH of 7 means. It is the situation where the concentrations of hydrogen ion and hydroxide ion are equal, both with the value of $10^{-7} \, \text{mol} \, l^{-1}$.

SUMMARY

It is likely that the numerical relationship between hydrogen and hydroxide ion concentrations and pH is by now quite clear. However, to summarise these connections, a table of hydrogen and hydroxide ion concentrations and pH values is given below. Before trying the following Problems, it is well worth using this table to practice taking any two of these three values to obtain the third.

NOTE: In this table, all the concentrations are given in full decimal form, eg 0.0001 as well as in scientific notation (or standard form), eg 10^{-4}. In practice, we would almost always use the latter form, except for values such as 0.1 and 0.01, as it is too easy to make mistakes handling numbers with a larger number of decimal places.

$[H^+(aq)]/mol\ l^{-1}$		$[OH^-(aq)]/mol\ l^{-1}$		pH
1	or 10^0	0.00000000000001	or 10^{-14}	0
0.1	or 10^{-1}	0.0000000000001	or 10^{-13}	1
0.01	or 10^{-2}	0.000000000001	or 10^{-12}	2
0.001	or 10^{-3}	0.00000000001	or 10^{-11}	3
0.0001	or 10^{-4}	0.0000000001	or 10^{-10}	4
0.00001	or 10^{-5}	0.000000001	or 10^{-9}	5
0.000001	or 10^{-6}	0.00000001	or 10^{-8}	6
0.0000001	or 10^{-7}	0.0000001	or 10^{-7}	7
0.00000001	or 10^{-8}	0.000001	or 10^{-6}	8
0.000000001	or 10^{-9}	0.00001	or 10^{-5}	9
0.0000000001	or 10^{-10}	0.0001	or 10^{-4}	10
0.00000000001	or 10^{-11}	0.001	or 10^{-3}	11
0.000000000001	or 10^{-12}	0.01	or 10^{-2}	12
0.0000000000001	or 10^{-13}	0.1	or 10^{-1}	13
0.00000000000001	or 10^{-14}	1	or 10^0	14

PROBLEMS

These problems combine the work of previous Worked Examples and problems, where, given any one of $[H^+(aq)]$, $[OH^-(aq)]$ or pH, the other two can be obtained.

13.21 A solution has a hydroxide ion concentration of $10^{-4}\ mol\ l^{-1}$. Calculate
    (*a*) the hydrogen ion concentration;
    (*b*) the pH.

13.22 A weakly acidic solution has a pH of 6. Calculate
    (*a*) the hydrogen ion concentration;
    (*b*) the hydroxide ion concentration.

13.23 An alkaline solution has a hydroxide ion concentration of $10^{-2}\ mol\ l^{-1}$. Calculate
    (*a*) the hydrogen ion concentration;
    (*b*) the pH.

13.24 A solution of ethanoic acid has a pH of 5. Calculate
- (a)   the hydrogen ion concentration;
- (b)   the hydroxide ion concentration.

13.25 A weakly acidic solution has a hydroxide ion concentration of 10^{-9} mol l^{-1}. Calculate
- (a)   the hydrogen ion concentration;
- (b)   the pH.

13.26 An alkali solution has a pH of 14. Calculate
- (a)   the hydrogen ion concentration;
- (b)   the hydroxide ion concentration.

13.27 A solution of an alkali has a hydroxide ion concentration of 10^{-3} mol l^{-1}. Calculate
- (a)   the hydrogen ion concentration;
- (b)   the pH.

13.28 A solution of sodium hydroxide has a pH of 13. Calculate
- (a)   the hydrogen ion concentration;
- (b)   the hydroxide ion concentration.

13.29 A hydrochloric acid solution has a hydroxide ion concentration of 10^{-13} mol l^{-1}. Calculate
- (a)   the hydrogen ion concentration;
- (b)   the pH.

13.30 A methanoic acid solution has a pH of 3. Calculate
- (a)   the hydrogen ion concentration;
- (b)   the hydroxide ion concentration.

ADDITIONAL THEORY; 2 The effect of dilution on pH

A 1 mol l^{-1} solution of hydrochloric acid has a pH of 0. If a sample of this solution is taken and **diluted by a factor of 10**, our new solution will be **a tenth** of the concentration of the original solution; that is it will be 0.1 mol l^{-1} and its pH value will be 1. The pH of the new solution is now 1. Typically this would be carried out experimentally by pipetting a 10 cm^3 sample of the original solution, emptying it into a 100 cm^3 standard flask and diluting the solution up the mark with water. (Or taking a 25 cm^3 sample and diluting it to 250 cm^3 etc.)

If we take a sample of this new solution and dilute it *further* by a factor of 10, we will have made a solution with a concentration of 0.01 mol l^{-1} and a pH value of 2. If we repeat this process several times, we end up with a range of concentrations and pH values as shown below.

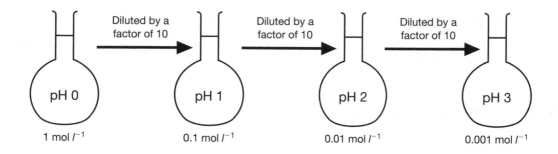

This process of repeated dilution (sometimes called 'serial dilution') by a factor of 10 gives us solutions with a range of pH numbers, increasing by 1 with each dilution.

This example started with an **acid** solution which was diluted to give solutions of pH 1, 2, 3 etc. A range of **alkaline** solutions, differing by 1 pH unit each time, can be similarly obtained. A 1 mol l^{-1} solution of sodium hydroxide will have a pH of 14. If such a solution is diluted by a factor of 10, a 0.1 mol l^{-1} solution, with pH 13 will be obtained. Successive dilutions by a factor of 10 each time, would give solutions with pH values 12, 11, 10 etc.

This can provide a useful method of obtaining a solution of a particular pH, starting from a more concentrated solution. This can be seen in the following Worked Examples.

Worked Example 13.8

Starting with 100 cm³ of 0.1 mol l^{-1} hydrochloric acid solution, describe how a laboratory technician could make up 250 cm³ of hydrochloric acid solution with a pH of 2.

From previous work, the original 0.1 mol l^{-1} (10^{-1} mol l^{-1}) solution has a pH of 1 and the new solution has to have a pH of 2. To get from pH 1 to pH 2 involves **dilution by a factor of 10**.

The volume of the new solution has to be 250 cm³, so the volume of the original solution required is 25 cm³. The technician would pipette 25 cm³ of the original solution into a 250 cm³ standard flask, add distilled water up to the mark and mix thoroughly. This solution would now be **a tenth** of the original concentration, ie 0.01 mol l^{-1}, and would have the required pH of 2.

Worked Example 13.9

A chemist has 250 cm^3 of 0.1 mol l^{-1} sodium hydroxide solution, but needs to make up 100 cm^3 of sodium hydroxide solution with a pH of 11. Describe how this new solution should be prepared.

The first step is to calculate the pH of the original solution, as in previous work.

The concentration of hydroxide ion is 0.1 mol l^{-1} (10^{-1} mol l^{-1}), so the hydrogen ion concentration must be 10^{-13} mol l^{-1}, and therefore the pH of the original solution is 13.

To get a pH of 12 would require dilution by a factor of 10. To get to a pH of 11 would require a **further dilution** by a factor of 10. That is, the chemist needs to dilute the original solution **by a factor of 100**. This could be done in two obvious ways.

The chemist needs 100 cm^3 of the new, diluted, solution, so 1 cm^3 of the original solution could be pipetted into a 100 cm^3 standard flask and the solution made up to the mark with distilled water. This would be carrying out the dilution **by a factor of 100** directly. The chemist now has the required volume of solution with a pH of 11.

The alternative would be to carry out a 'serial dilution' as described earlier.

The chemist could take 10 cm^3 of the original alkali solution using a pipette and pour it into a 100 cm^3 standard flask, making the solution up to the mark with distilled water. The solution has now been diluted **by a factor of 10** and has a pH of 12. The chemist would then take 10 cm^3 of this diluted solution, empty it into a new standard flask and make it up to 100 cm^3 with distilled water. The solution has now been diluted by a **further factor of 10**, that is the original solution has been diluted **by a factor of 100**. The chemist now has the required volume of solution with a pH of 11.

PROBLEMS

These problems are of the type shown in Worked Examples 13.8 and 13.9.

13.31 A pupil has 100 cm^3 of 1 mol l^{-1} hydrochloric acid solution.
  (*a*) What is the pH of this solution?
  (*b*) Describe how the pupil could use this solution to make up 100 cm^3 of hydrochloric acid solution with a pH of 1.

13.32 A pupil has 100 cm^3 of 0.01 mol l^{-1} nitric acid solution.
  (*a*) What is the pH of this solution?
  (*b*) Describe how the pupil could use this solution to make up 50 cm^3 of nitric acid solution with a pH of 3.

13.33 A technician has $100 \, cm^3$ of $0.1 \, mol \, l^{-1}$ hydrochloric acid solution.
   (*a*)   What is the pH of this solution?
   (*b*)   Describe how the technician could use this solution to make up $250 \, cm^3$ of hydrochloric acid solution with a pH of 2.

13.34 A technician has $10 \, cm^3$ of $0.1 \, mol \, l^{-1}$ nitric acid solution.
   (*a*)   What is the pH of this solution?
   (*b*)   Describe how the technician could use this solution to make up $500 \, cm^3$ of nitric acid solution with a pH of 3.

13.35 A pupil has $50 \, cm^3$ of $1 \, mol \, l^{-1}$ hydrochloric acid solution.
   (*a*)   What is the pH of this solution?
   (*b*)   Describe how the pupil could use this solution to make up 1 litre of hydrochloric acid solution with a pH of 2.

13.36 A pupil has $100 \, cm^3$ of $1 \, mol \, l^{-1}$ sodium hydroxide solution, NaOH.
   (*a*)   What is the pH of this solution?
   (*b*)   Describe how the pupil could use this solution to make up $50 \, cm^3$ of sodium hydroxide solution with a pH of 13.

13.37 A pupil has $100 \, cm^3$ of $0.1 \, mol \, l^{-1}$ potassium hydroxide solution, KOH.
   (*a*)   What is the pH of this solution?
   (*b*)   Describe how the pupil could use this solution to make up $100 \, cm^3$ of potassium hydroxide solution with a pH of 12.

13.38 A technician has $100 \, cm^3$ of $0.01 \, mol \, l^{-1}$ potassium hydroxide solution, KOH.
   (*a*)   What is the pH of this solution?
   (*b*)   Describe how the technician could use this solution to make up $250 \, cm^3$ of potassium hydroxide solution with a pH of 11.

13.39 A pupil has $50 \, cm^3$ of $1 \, mol \, l^{-1}$ sodium hydroxide solution.
   (*a*)   What is the pH of this solution?
   (*b*)   Describe how the pupil could use this solution to make up 1 litre of sodium hydroxide solution with a pH of 12.

13.40 A technician has $10 \, cm^3$ of $0.01 \, mol \, l^{-1}$ sodium hydroxide solution.
   (*a*)   What is the pH of this solution?
   (*b*)   Describe how the technician could use this solution to make up $500 \, cm^3$ of sodium hydroxide solution with a pH of 11.

Calculations from Equations – 7: Involving Redox Titrations

This chapter involves calculations from equations where the chemical reaction is a **redox reaction**; that is, one where one reactant is being **reduced** and the other is being **oxidised**. The former reactant **gains** electrons; the latter **loses** electrons.

In acid-alkali reactions, the concentration of one solution can be obtained by titration with another of known concentration, using an indicator which changes colour at the 'end-point' – the point where one substance has *just* neutralised the other.

Redox reactions can also be used to find the concentration of an 'unknown' solution. Sometimes the redox reaction is 'self-indicating', meaning that one of the reactants changes colour sharply at the end-point. Common reactants of this type include the permanganate ion, $MnO_4^-(aq)$ and the dichromate ion, $Cr_2O_7^{2-}(aq)$, which, in acid solution, can undergo the following reductions, with the colour changes shown. (These ion-electron equations are included on p. 11 of the SQA Data Book.)

$$MnO_4^-(aq) \ + \ 8H^+(aq) \ + \ 5e^- \ \longrightarrow \ Mn^{2+}(aq) \ + \ 4H_2O(l)$$
PURPLE COLOURLESS

$$Cr_2O_7^{2-}(aq) \ + \ 14H^+(aq) + 6e^- \ \longrightarrow \ Cr^{3+}(aq) \ + \ 7H_2O(l)$$
ORANGE GREEN

Other redox reactions will require the presence of an added indicator which will change colour at the end-point. A common example is starch which forms a purple-black colour in the presence of iodine, I_2.

In calculations involving redox reactions, it is not always necessary to have a full balanced equation for the overall reaction; there is usually only a need to balance the number of electrons being gained in the reduction with the number of electrons being lost in the oxidation. This can be seen in the following Worked Example.

Worked Example 14.1

The ion-electron equations below represent the oxidation and reduction reactions taking place when permanganate ions, MnO_4^-, in acid solution, react with iodide ions.

RED: $MnO_4^- + 8H^+ + 5e^- \longrightarrow Mn^{2+} + 4H_2O$

OX: $2I^- \longrightarrow I_2 + 2e^-$

20 cm³ of a 0.2 mol l^{-1} solution of iodide ions is titrated with a 0.064 mol l^{-1} solution of permanganate ions, in acid solution. What volume of the permanganate solution will be required to react exactly with all the iodide ions?

This problem is actually not different from those done previously, except that the balanced equation for the overrall redox reaction has not been given. Using the ion-electron equations given, it can be seen that the reduction reaction ('RED') involves the gain of 5 electrons, while the oxidation reaction ('OX') involves the loss of 2 electrons. Since the number of electrons being gained must be the same as that being lost, the balanced equation is obtained by multiplying the above equations as below:

$2 \times$ RED: $2 \times (MnO_4^- + 8H^+ + 5e^- \longrightarrow Mn^{2+} + 4H_2O)$

$5 \times$ OX: $5 \times (2I^- \longrightarrow I_2 + 2e^-)$

If the reduction is multiplied throughout by 2, we now have a **gain of 10 electrons**. The multiplication of the oxidation by 5 gives a **loss of 10 electrons**. That is, the two 'half-reactions' are now balanced. The full balanced equation can be obtained by multiplying out the reduction and oxidation reactions, adding them together and cancelling out species common to each side. However, for the purpose of this problem, that is not necessary. What is important to see is that the combined reduction and oxidation reactions tell us that:

2 mol of MnO_4^- reacts with 10 mol of I^-

Or, more simply, 1 mol of MnO_4^- reacts with 5 mol of I^-.
Thus, we have established 'Step 2: Mole Statement', in which the number of moles of the species we are asked about and told about in the problem are connected.
We proceed as in previous examples.

Step 3: Calculation of 'Known' Moles
'Known' substance is I^-

$$\text{number of moles} = \text{concentration} \times \text{number of litres}$$
$$= 0.2 \times 0.02$$
$$= 0.004 \text{ mol of } I^-$$

Step 4: Calculation of 'Unknown' Moles
From the Mole Statement:

$$1 \text{ mol of } MnO_4^- \text{ reacts with 5 mol of } I^-$$

Reversing to put MnO_4^- on the right hand side, we have:

$$5 \quad \text{mol of } I^- \text{ reacts with} \quad 1 \quad \text{mol of } MnO_4^-$$

$$1 \quad \text{mol of } I^- \text{ reacts with} \quad \frac{1}{5} \quad \text{mol of } MnO_4^-$$

$$0.004 \quad \text{mol of } I^- \text{ reacts with} \quad \frac{0.004}{5} \quad \text{mol of } MnO_4^-$$

$$= 0.0008 \text{ mol of } MnO_4^-$$

Step 5: Finishing Off

$$\text{volume (in litres)} = \frac{\text{number of moles}}{\text{concentration}}$$
$$= \frac{0.0008}{0.064}$$
$$= 0.0125 \, l \, (12.5 \text{ cm}^3)$$

PROBLEMS

These problems involve redox reactions of the type illustrated in Worked Example 14.1.

14.1
$$Cr_2O_7^{2-} + 14H^+ + 6e^- \longrightarrow 2Cr^{3+} + 7H_2O$$
$$SO_3^{2-} + H_2O \longrightarrow SO_4^{2-} + 2H^+ + 2e^-$$

The above ion-electron equations represent the reduction and oxidation reactions which take place when a solution of dichromate ions, $Cr_2O_7^{2-}$, in acid solution react with sulphite ions, SO_3^{2-}.

What volume of a $0.05 \text{ mol } l^{-1}$ solution of dichromate ions would react with 30 cm^3 of a $0.25 \text{ mol } l^{-1}$ solution of sulphite ions?

14.2
$$MnO_4^- + 8H^+ + 5e^- \longrightarrow Mn^{2+} + 4H_2O$$
$$2Cl^- \longrightarrow Cl_2 + 2e^-$$

What volume of a $0.24 \text{ mol } l^{-1}$ solution of acidified permanganate ions would exactly react with 120 cm^3 of a $0.16 \text{ mol } l^{-1}$ solution of chloride ions?

14.3
$$I_2 + 2e^- \longrightarrow 2I^-$$
$$2S_2O_3^{2-} \longrightarrow S_4O_6^{2-} + 2e^-$$

25 cm^3 of a $0.05 \text{ mol } l^{-1}$ solution of thiosulphate ions, $S_2O_3^{2-}$, reacts exactly with 10 cm^3 of a solution of I_2. What concentration is the I_2 solution?

14.4

$$Fe^{3+} + e^- \longrightarrow Fe^{2+}$$
$$SO_3^{2-} + H_2O \longrightarrow SO_4^{2-} + 2H^+ + 2e^-$$

25 cm^3 of a solution containing $0.016 \text{ mol } l^{-1}$ sulphite ions is titrated with $0.02 \text{ mol } l^{-1}$ Fe^{3+} solution. What volume of the Fe^{3+} solution would be required to obtain the exact end-point of this titration?

14.5

$$2BrO_3^- + 12H^+ + 10e^- \longrightarrow Br_2 + 6H_2O$$
$$2I^- \longrightarrow I_2 + 2e^-$$

32 cm^3 of a $0.0125 \text{ mol } l^{-1}$ solution of bromate ions, BrO_3^-, exactly oxidises 25 cm^3 of a solution of iodide ions. What is the concentration of the iodide ion solution?

14.6

$$2HClO + 2H^+ + 2e^- \longrightarrow Cl_2 + 2H_2O$$
$$2I^- \longrightarrow I_2 + 2e^-$$

25 cm^3 of a $0.02 \text{ mol } l^{-1}$ solution of hypochlorous acid, $HClO$, is exactly reduced by 10 cm^3 of a solution of iodide ions. What is the concentration of the iodide ion solution?

14.7

$$Fe^{3+} + e^- \longrightarrow Fe^{2+}$$
$$SO_3^{2-} + H_2O \longrightarrow SO_4^{2-} + 2H^+ + 2e^-$$

12.8 cm^3 of a $0.12 \text{ mol } l^{-1}$ solution of sulphite ions is exactly oxidised by 7.68 cm^3 of a solution of iron(III) ions. What is the concentration of the iron(III) ions present?

14.8

$$Cr_2O_7^{2-} + 14H^+ + 6e^- \longrightarrow Cr^{3+} + 7H_2O$$
$$Fe^{2+} \longrightarrow Fe^{3+} + e^-$$

40 cm^3 of a solution containing acidified dichromate ions with a concentration of $0.015 \text{ mol } l^{-1}$ is titrated with $0.2 \text{ mol } l^{-1}$ Fe^{2+} solution. What volume of the Fe^{2+} solution would be required to obtain the exact end-point of this titration?

14.9

$$(COOH)_2 \longrightarrow 2CO_2 + 2H^+ + 2e^-$$
$$MnO_4^- + 8H^+ + 5e^- \longrightarrow Mn^{2+} + 4H_2O$$

A $0.01 \text{ mol } l^{-1}$ solution of oxalic acid, $(COOH)_2$, is titrated against 16 cm^3 of an acidified solution containing $0.005 \text{ mol } l^{-1}$ permanganate ions until the end-point is reached. What volume of oxalic acid solution must have reacted?

$$H_2O_2 + 2H^+ + 2e^- \longrightarrow 2H_2O$$
$$Fe^{2+} \longrightarrow Fe^{3+} + e^-$$

14.10 10 cm^3 of a $2.4 \text{ mol } l^{-1}$ solution of hydrogen peroxide, H_2O_2, is titrated with a solution containing $1.25 \text{ mol } l^{-1}$ of Fe^{2+}. What volume of the Fe^{2+} solution will be required to react exactly with the hydrogen peroxide?

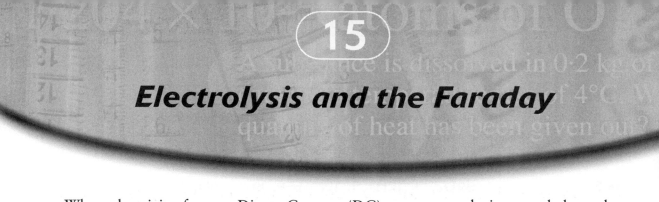
Electrolysis and the Faraday

When electricity from a Direct Current (DC) power supply is passed through a solution of an ionic compound, or a melted ionic compound, reactions take place at the electrodes. A simple example of this is when electricity is passed through copper(II) chloride solution as shown below.

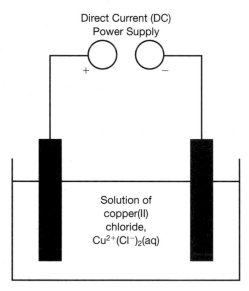

Direct Current (DC)
Power Supply

+   −

Solution of
copper(II)
chloride,
$Cu^{2+}(Cl^-)_2(aq)$

Electrons flowing from the negative terminal of the power supply are received by the Cu^{2+} ions to form copper metal which sticks to or 'plates' to the electrode. The ion equation for this **reduction** reaction is given below.

$$Cu^{2+}(aq) + 2e^- \longrightarrow Cu\,(s)$$

At the positive electrode, chloride ions lose electrons to form chlorine gas. This **oxidation** reaction is described by the ion-electron equation below.

$$2Cl^-(aq) \longrightarrow Cl_2(g) + 2e^-$$

The overall process is called **electrolysis.**
In an electrolysis, it should be obvious that the more electrical charge which flows round the circuit, the more material will be produced at each electrode.
In a circuit, the speed at which electrical charge flows is the **current**, symbol *I*. Current is measured in units of **Amps** (short for Amperes), symbol **A**.

Electrical **charge** has the symbol Q. It is measured in units of **Coulombs**, symbol **C**. The **time** the current has been flowing for has the symbol t. It is measured in units of **seconds**, symbol **s**.

The total charge which flows in the circuit is obtained by the equation below.

$$Q = It$$

In an electrical circuit, the charge is carried through the wires by electrons. A particularly important quantity of electrical charge is the charge on 1 mol of electrons. This has the value of 96,500 C and is known as the Faraday, symbol F.

1 Faraday(1F) = 96,500 C = the charge on 1 mol of electrons

This value can be calculated experimentally, as described in Part A, next.
In Part B, the known value of the Faraday is used to calculate the mass of metal or the mass or volume of gas produced in an electrolysis.
Part C involves more complex problems involving the Faraday.

PART A: Calculating the Faraday experimentally.

In this part, two methods for obtaining the value for the Faraday by experiment are described.
In Worked Example 15.1, an electrolysis of a metal ion solution is carried out and the mass of metal deposited on the negative electrode is used to calculate the Faraday.
In Worked Example 15.2, the volume of gas given off at an electrode in an electrolysis is measured, and, knowing the molar volume of the gas, the Faraday can be calculated.

Worked Example 15.1

In this first method, a solution containing the ions of an unreactive metal is electrolysed using the apparatus shown below. In this case, the solution is copper(II) sulphate.

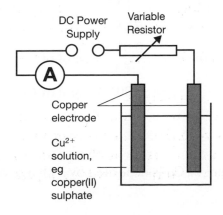

During the electrolysis, copper ions are reduced to copper metal at the negative electrode, according to the ion-electron equation below

$$Cu^{2+}(aq) + 2e^- \longrightarrow Cu(s)$$

In this experiment, the mass of the negative electrode increases due to copper being deposited. The increase in mass is obtained by weighing the electrode before the current is switched on and reweighing it after switching off. The current is kept at a known, constant, value by adjusting the variable resistor (as in a dimmer switch or volume control) as necessary. The length of time that the current has been flowing is also recorded.

Typical Results and Specimen Calculation

Current: = 0.2 A
Time: = 20 minutes = 1200 s
Final mass of electrode: = 24.57 g
Initial mass of electrode: = 24.49 g

The quantity of electrical charge is calculated:

$$Q = It$$
$$= 0.2 \times 1200$$
$$= 240 \text{ C}$$

The increase in mass is obtained by subtraction:

$$24.57 - 24.49 = 0.08 \text{ g}$$

This mass of copper is converted into a number of moles.

$$63.5 \text{ g} = 1 \qquad \text{mol of Cu}$$
$$1 \text{ g} = \frac{1}{63.5} \qquad \text{mol of Cu}$$
$$0.08 \text{ g} = \frac{0.08 \times 1}{63.5} \text{ mol of Cu}$$
$$= 0.00126 \text{ mol of Cu (rounded)}$$

So,

0.00126 mol of Cu is deposited by a charge of 240 C

Therefore,

$$\text{1 mol of Cu is deposited by } \frac{240}{0.00126} \text{ C}$$

$$= 190476 \text{ C}$$

The depositing of Cu is represented by the ion-electron equation:

$$Cu^{2+} + 2e^- \longrightarrow Cu$$
$$\text{2 mol} \qquad \text{1 mol}$$

This tells us that 1 mol of Cu is deposited by 2 mol of electrons flowing in the circuit.

So the charge calculated above (190476 C) which would deposit **1 mol of Cu** must be the charge on **2 mol of e⁻**.

$$\text{So the calculated charge on 1 mol of electrons is } \frac{190476}{2} \text{ C}$$

$$= 95200 \text{ C (rounded)}$$

This is close to the accepted value for the Faraday of 96500 C. The difference between the calculated and the accepted values can be explained by experimental error. The accepted value has been obtained under rigorous scientific conditions where error is reduced to a minimum.

PROBLEMS

In Problems 15.1 to 15.10, an aqueous solution of a metal ion solution is being electrolysed where a metal is deposited on the negative electrode.

In each of the following problems, calculate the value of the Faraday which would be obtained from the experimental information given.

Note that the number of moles of electrons required to deposit 1 mol of the metal is obtained from the number of positive charges on the metal ion. Where a metal can form ions with different charges, the charge is stated in the name of the compound; for example, copper(II) sulphate contains the Cu^{2+} ion. It would therefore need 2 mol of electrons to deposit 1 mol of copper metal.

15.1 A solution of copper(II) sulphate was electrolysed for 30 minutes using a current of 0.15 A. A mass of 0.089 g of copper was deposited on the negative electrode.

15.2 A solution of nickel(II) nitrate was electrolysed for 40 minutes using a current of 0.2 A. A mass of 0.146 g of nickel was deposited on the negative electrode.

15.3 A solution of silver(I) nitrate was electrolysed for 25 minutes using a current of 0.2 A. A mass of 0.335 g of silver was deposited on the negative electrode.

15.4 A solution of copper(II) chloride was electrolysed for 45 minutes using a current of 0.2 A. A mass of 0.178 g of copper was deposited on the negative electrode.

15.5 A solution of nickel(II) sulphate was electrolysed for 20 minutes using a current of 0.12 A. A mass of 0.0438 g of nickel was deposited on the negative electrode.

15.6 A solution of silver(I) nitrate was electrolysed for 30 minutes using a current of 0.1 A. A mass of 0.201 g of silver was deposited on the negative electrode.

15.7 A solution of copper(II) nitrate was electrolysed for 20 minutes using a current of 0.18 A. A mass of 0.0711 g of copper was deposited on the negative electrode.

15.8 A solution of silver(I) nitrate was electrolysed for 40 minutes using a current of 0.15 A. A mass of 0.402 g of silver was deposited on the negative electrode.

15.9 A solution of nickel(II) chloride was electrolysed for 25 minutes using a current of 0.24 A. A mass of 0.111 g of nickel was deposited on the negative electrode.

15.10 A solution of copper(II) sulphate was electrolysed for 1 hour using a current of 0.24 A. A mass of 0.284 g of copper was deposited on the negative electrode.

Worked Example 15.2

This method is similar to the previous one described in Worked Example 15.1, except that in this case the volume of a gas being given off in the electrolysis is measured. In this case, an acid solution (containing a large number of hydrogen ions, $H^+(aq)$), is electrolysed and hydrogen gas is collected at the negative electrode. This is represented by the reduction reaction below:

$$2H^+(aq) + 2e^- \rightarrow H_2(g)$$

Typical apparatus is shown below.

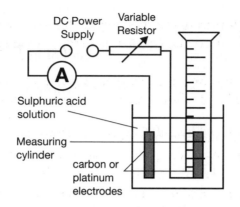

As in Worked Example 15.1, the current is kept at a known, constant, value by adjusting the variable resistor as necessary. The time the current has been flowing is noted and the volume of gas produced is measured.

Typical Results and Specimen Calculation

Current: = 0.2 A

Time: = 20 minutes = 1200 s

Volume of H_2 gas collected: = 30.25 cm^3

The quantity of electrical charge is calculated as before:

$$Q = It$$
$$= 0.2 \times 1200$$
$$= \mathbf{240\,C}$$

We have the volume of hydrogen gas, but to get the **number of moles** of hydrogen given off we need to know its **molar volume** (ie the volume of 1 mol.) Under the conditions of this experiment, the molar volume of hydrogen is 24 litres.

Firstly the volume of 30.25 cm^3 is converted to litres:

$$30.25 \text{ cm}^3 = \frac{30.25}{1000} \text{ litres}$$
$$= \mathbf{0.03025 \text{ litres}}$$

This volume is then converted to a number of moles using the known molar volume of the gas (24 litres).

$$24 \quad \text{litres} = \quad 1 \quad \text{mol of hydrogen}$$

$$1 \quad \text{litre} = \quad \frac{1}{24} \quad \text{mol of hydrogen}$$

$$0.03025 \text{ litre} = \frac{0.03025 \times 1}{24} \quad \text{mol of hydrogen}$$

$$= 0.00126 \text{ mol of hydrogen (rounded)}$$

So 0.00126 mol of H$_2$ is produced by 240 C of charge flowing

0.00126 mol of H$_2$(g) is deposited by a charge of 240 C

Therefore,

1 mol of H$_2$(g) is deposited by a charge of $\dfrac{240}{0.00126}$ **C**

$$= 190476 \text{ C}$$

The production of hydrogen gas at the negative electrode is represented by the ion-electron equation:

$$2\text{H}^+(\text{aq}) + 2\text{e}^- \longrightarrow \text{H}_2(\text{aq})$$
$$\text{2 mol} \qquad \text{1 mol}$$

This tells us that **1 mol of H$_2$(g)** is given off by **2 mol of electrons** flowing in the circuit.

So the charge calculated above (190476 C) which resulted in **1 mol of H$_2$(g)** being given off must be the charge on **2 mol of electrons**.

So the charge on **1 mol of electrons** is $\dfrac{190476}{2}$ C

$$= 95200 \text{ C (rounded)}$$

This is the same value as that calculated in Worked Example 15.1 and is reasonably close to the accepted value for the Faraday of 96500 C, taking into account experimental error.

NOTE: In the electrolysis of *any* aqueous solution, there is the possibility of hydrogen being produced at the negative electrode and/or oxygen being produced at the positive electrode. The ion-electron equations describing these electrode reactions are given below. These equations are given in the SQA Data Book for Higher and Advanced Higher, August 1999 Edition, page 11, under the title 'Electrolysis of Water'. These are also given in Appendix 2 on page 142 of this book and are repeated below.

Reduction reactions at the negative electrode.

$$2H_2O(l) + 2e^- \longrightarrow H_2(g) + 2OH^-(aq)$$

$$2H^+(aq) + 2e^- \longrightarrow H_2(g)$$

The **first** equation would apply in a **non-acid** aqueous solution, ie a neutral or alkaline solution, where there were plenty of water molecules, $H_2O(l)$, but very few hydrogen ions, $H^+(aq)$.

The **second** equation would apply in an **acid** solution where there were plenty of hydrogen ions, $H^+(aq)$.

In **both cases,** the important point to note is that the relationship below applies:

2 mol of e^- flowing in the circuit will cause **1 mol of $H_2(g)$** to be given off.

Oxidation reactions at the positive electrode.

$$2H_2O(l) \longrightarrow O_2(g) \quad + \quad 4H^+(aq) + 4e^-$$

$$4OH^-(aq) \longrightarrow 2H_2O(l) + O_2(g) \quad + \quad 4e^-$$

The **first** equation would apply in a **non-alkaline** aqueous solution, ie a neutral or acid solution, where there were plenty of water molecules, $H_2O(l)$, but very few hydroxide ions, $OH^-(aq)$.

The **second** equation would apply in an **alkaline** solution where there were plenty of hydroxide ions, $OH^-(aq)$.

In **both cases,** the important point to note is that the relationship below applies:

4 mol of e^- flowing in the circuit will cause **1 mol of $O_2(g)$** to be given off.

PROBLEMS

In problems 15.11 to 15.15, the value of the Faraday has to be calculated by measuring the volume of **hydrogen gas** being given off at the **negative** electrode during electrolysis of an aqueous solution. With reference to the ion-electron equations, the important relationship is that:

At the negative electrode, 2 mol of electrons produce 1 mol of $H_2(g)$.

In each of the following problems, calculate the value of the Faraday which would be obtained from the experimental information given.

15.11 A current of 0.2 A is passed through an acid solution for 20 minutes, during which time 24.9 cm³ of hydrogen gas was given off at the negative electrode. The molar volume of hydrogen under these conditions was 20 litres.

15.12 A current of 0.15 A is passed through an acid solution for 1 hour, during which time 61.6 cm³ of hydrogen gas was given off at the negative electrode. The molar volume of hydrogen under these conditions was 22 litres.

15.13 A current of 0.25 A is passed through an acid solution for 40 minutes, during which time 74.6 cm³ of hydrogen gas was given off at the negative electrode. The molar volume of hydrogen under these conditions was 24 litres.

15.14 A current of 0.1 A is passed through an acid solution for 30 minutes, during which time 23.3 cm³ of hydrogen gas was given off at the negative electrode. The molar volume of hydrogen under these conditions was 25 litres.

15.15 A current of 0.12 A is passed through an acid solution for 25 minutes, during which time 19.6 cm³ of hydrogen gas was given off at the negative electrode. The molar volume of hydrogen under these conditions was 21 litres.

PROBLEMS

In problems 15.16 to 15.20, the value of the Faraday has to be calculated by measuring the volume of **oxygen gas** being given off at the **positive** electrode during electrolysis of an aqueous solution. With reference to the ion-electron equations referred to on p??, the important relationship is that:

At the positive electrode, 4 mol of electrons produce 1 mol of $O_2(g)$.

In each of the following problems, calculate the value of the Faraday which would be obtained from the experimental information given.

15.16 A current of 0.15 A is passed through an aqueous solution for 40 minutes, during which time 21.5 cm³ of oxygen gas was given off at the positive electrode. The molar volume of oxygen under these conditions was 23 litres.

15.17 A current of 0.25 A is passed through an aqueous solution for 1 hour, during which time 44.3 cm³ of oxygen gas was given off at the positive electrode. The molar volume of oxygen under these conditions was 19 litres.

15.18 A current of 0.18 A is passed through an aqueous solution for 30 minutes, during which time 20.1 cm³ of oxygen gas was given off at the positive electrode. The molar volume of oxygen under these conditions was 24 litres.

15.19 A current of 0.2 A is passed through an aqueous solution for 25 minutes, during which time 18.6 cm^3 of oxygen gas was given off at the positive electrode. The molar volume of oxygen under these conditions was 24 litres.

15.20 A current of 0.1 A is passed through an aqueous solution for 50 minutes, during which time 16.3 cm^3 of oxygen gas was given off at the positive electrode. The molar volume of oxygen under these conditions was 21 litres.

PART B: Calculating the amount of substance produced at an electrode during electrolysis using the Faraday.

In this part, the value of the Faraday, 96500 C, is used to calculate the mass of metal deposited at the negative electrode, or the volume or mass of gas given off at either electrode.

Worked Example 15.3

What mass of nickel is deposited in the electrolysis of nickel(II) sulphate solution if a current of 0.4 A is passed for 120 minutes?

The total charge passed is calculated:

$$Q = It$$
$$= 0.4 \times 120 \times 60 \text{ (Note: time is in seconds.)}$$
$$= 2880 \text{ C}$$

We then calculate the number of moles of electrons that this charge represents:

$$96500 \text{ C} \quad \text{is the charge on} \quad 1 \quad \text{mol of electrons}$$

$$1 \text{ C} \quad \text{is the charge on} \quad \frac{1}{96\,500} \quad \text{mol of electrons}$$

$$2880 \text{ C} \quad \text{is the charge on} \quad \frac{2880}{96\,500} \quad \text{mol of electrons}$$

$$= 0.02984 \text{ mol of electrons (rounded)}$$

We now write the ion-electron equation to relate the number of moles of electrons with the number of moles of nickel deposited. The compound being electrolysed is nickel(II) sulphate, so the nickel ion present is $Ni^{2+}(aq)$ and the equation for the deposition of $Ni(s)$ metal must be:

$$Ni^{2+} + 2e^- \longrightarrow Ni$$

This equation tells us that:

2	mol of electrons deposits	1	mol of Ni
1	mol of electrons deposits	$\frac{1}{2}$	mol of Ni
0.02984	mol of electrons deposits	$0.02984 \times \frac{1}{2}$	mol of Ni

$$= 0.01492 \text{ mol of Ni}$$

The number of moles of nickel deposited has been calculated, but the problem asks for the mass of nickel. This is the final part of the calculation.

$$1 \quad \text{mol of Ni} = 58.7 \quad \text{g}$$
$$0.01492 \text{ mol of Ni} = 0.01492 \times 58.7 \text{ g}$$
$$= 0.876 \text{ g (rounded to 3 significant figures)}$$

PROBLEMS

The following problems are of the type illustrated by Worked Example 15.3, involving the mass of metal deposited at the negative electrode, or the mass of gas given off at either electrode.

Where a metal is being deposited, whether from a solution, or from a molten compound, the number of moles of electrons involved in depositing 1 mol of metal can be obtained from the charge on the metal ion. For metals in Groups 1, 2 and 3 of the Periodic Table, the charges are $+$, $2+$ and $3+$. Where a metal outside these Groups is involved, the charge on the metal ion is given in the name of the compound; eg. silver(I) nitrate has the Ag^+ ion, iron(III) oxide has the Fe^{3+} ion, etc.

In the problems below, the ion-electron equations are only given where they may be unfamiliar.

Although the later problems (15.30–15.35) involve large currents used in industrial electrolyses, the same method should be used as for the earlier 'laboratory scale' problems.

15.21 In the electrolysis of copper(II) sulphate solution, a current of 0.1 A flowed for 60 minutes. Calculate the mass of copper which was deposited at the negative electrode.

15.22 In the electrolysis of a molten tin(IV) compound, a current of 0.2 A flowed for 15 minutes. Calculate the mass of tin deposited.

15.23 In the electrolysis of silver(I) nitrate solution, a current of 0.15 A flowed for 30 minutes. Calculate the mass of silver deposited.

15.24 A dilute solution of sulphuric acid is electrolysed by passing a current of 0.25 A through it for 20 minutes. The ion-electron equation for the reaction taking place at the negative electrode is:

$$2H^+ (aq) + 2e^- \longrightarrow H_2 (g)$$

Calculate the mass of hydrogen gas given off at this electrode during the electrolysis.

15.25 Molten lead(II) bromide is electrolysed using a current of 0.4 A for two hours. Calculate the mass of lead produced at the negative electrode.

15.26 The electrode reactions taking place in the electrolysis of sodium hydroxide solution are:

positive clectrode: $4OH^- (aq) \longrightarrow 2H_2O (l) + O_2 (g) + 4e^-$
negative electrode: $2H_2O + 2e^- \longrightarrow H_2 + 2OH^-$

Sodium hydroxide solution is electrolysed using a current of 0.15 A for 30 minutes. Calculate the mass of *(a)* hydrogen and *(b)* oxygen evolved at the electrodes.

15.27 In the electrolysis of molten aluminium oxide, a current of 0.25 A flowed for two hours. Calculate the mass of aluminium deposited.

15.28 In the electrolysis of dilute nitric acid, the electrode reactions are:

positive electrode: $2H_2O (l) \longrightarrow O_2 (g) + 4H^+ (aq) + 4e^-$
negative electrode: $2H^+ (aq) + 2e^- \longrightarrow H_2 (g)$

Dilute nitric acid is electrolysed using a current of 0.15 A for 40 minutes. Calculate the mass of *(a)* hydrogen and *(b)* oxygen evolved.

15.29 Copper is purified by electrolysing copper(II) sulphate solution with a positive electrode of impure copper and a negative electrode of pure copper. As the electrolysis takes place, the impure copper dissolves in the solution and pure copper is deposited on the negative electrode. If a current of 50 A was passed for ten hours in such an electrolysis, what mass of pure copper would be deposited?

15.30 In the industrial manufacture of aluminium, molten aluminium oxide is electrolysed using a current of 10^5 A. Calculate the mass, in kg, of aluminium produced in a period of 24 hours by this process.

15.31 Sodium is manufactured by the electrolysis of molten sodium chloride using a current of 2.5×10^4 A. Calculate the mass, in kg, of sodium produced in a two hour period.

15.32 A current of 500 A is used to nickel plate metal objects by the electrolysis of a solution of a nickel(II) salt. Calculate the mass, in kg, of nickel which would be deposited every 24 hours by this process.

15.33 Magnesium is produced electrolytically using a current of 2×10^5 A through a molten magnesium compound. What mass of magnesium, in kg, would be produced during an eight hour period by this process?

15.34 Chromium can be plated on metal objects by the electrolysis of a chromium(III) solution. What mass of chromium would be plated on the objects if a current of 2×10^3 A were applied for one hour?

15.35 An electrolytic smelter uses a current of 1.4×10^5 A to obtain aluminium from its molten oxide. What mass of aluminium, in kg, would be produced in eight hours of this smelter's operation?

Worked Example 15.4

What volume of hydrogen would be given off at the negative electrode in the electrolysis of an aqueous solution if a current of 0.25 A flowed for 2 hours, and if the molar volume of hydrogen under the conditions of measurement was 24 litres?

Before starting the calculation, note that, as previously mentioned, the possible ion-electron equations describing hydrogen being given off are:

Reduction reactions at the negative electrode.

$$2H_2O(l) + 2e^- \longrightarrow H_2(g) + 2OH^-(aq)$$
$$2H^+(aq) + 2e^- \longrightarrow H_2(g)$$

Note again that these equations are given in the SQA Data Book, and are included as Appendix 2 on page 142 to this book.

The important point to note is that, in both equations, as highlighted by the **bold** type: **2 mol of e^-** flowing causes **1 mol of $H_2(g)$** to be given off.

Proceeding with the calculation,

$$Q = It$$
$$= 0.25 \times 2 \times 60 \times 60 \text{ C}$$
$$= 1800 \text{ C}$$

Using the value of the Faraday, 96500 C,

$$96500 \text{ C} = 1 \text{ mol of e}^-$$
$$1 \text{ C} = \frac{1}{96500} \text{ mol of e}^-$$
$$1800 \text{ C} = \frac{1 \times 1800}{96500} \text{ mol of e}^-$$
$$= 0.01865 \text{ mol of e}^- \text{ (rounded)}$$

Repeating the information from the possible ion electron equations, we have:

2 mol of e$^-$ flowing causes 1 mol of H_2(g) to be given off.

So

1 mol of e$^-$ flowing causes $\dfrac{1}{2}$ mol of H_2(g) to be given off.

0.01865 mol of e$^-$ flowing causes $\dfrac{1 \times 0.01865}{2}$ mol of H_2(g) to be given off.

$$= 0.009325 \text{ mol of } H_2(g)$$

We are told that the molar volume of hydrogen is 24 litres, so

1 mol of H_2(g) has a volume of 24 litres

0.009325 mole of H_2(g) has a volume of 0.009325×24 litres

$$= 0.224 \text{ litres (rounded)}$$
$$\text{(or 22.4 cm}^3\text{)}$$

PROBLEMS

These problems are of the type shown by Worked Example 15.4 in which a gas is given off at an electrode and where its molar volume under the conditions of measurement is known.

NOTE: The only gases considered in these problems are hydrogen, oxygen and chlorine. The possible equations for hydrogen or oxygen being given off are given on page 126 of this Chapter, in Appendix 2 on page 142 and on p11 of the SQA Data Book. The ion-electron equation for the evolution of chlorine is:

$$2Cl^-(aq) \longrightarrow Cl_2(g) + 2e^-$$

This **oxidation** reaction is the reverse of the **reduction** equation given on p11 of the SQA Data Book.

15.36 In the electrolysis of an aqueous solution, a current of 0.2 A was passed for 20 minutes. What volume of hydrogen gas would be given off at the negative electrode if the molar volume of hydrogen is 25 litres?

15.37 In the electrolysis of a chloride ion solution, a current of 0.4 A was passed for 1 hour. What volume of chlorine gas would be given off at the positive electrode if the molar volume of chlorine is 32 litres? (Assume that chlorine is the only gas evolved at this electrode.)

15.38 In the electrolysis of an aqueous solution, a current of 0.25 A was passed for 10 minutes. What volume of oxygen gas would be given off at the positive electrode if the molar volume of oxygen is 40 litres?

15.39 In the electrolysis of an acid solution, a current of 0.1 A was passed for 25 minutes. What volume of hydrogen gas would be given off at the negative electrode if the molar volume of hydrogen is 30 litres?

15.40 In the electrolysis of an aqueous solution, a current of 0.2 A was passed for 2 hours. What volume of oxygen gas would be given off at the positive electrode if the molar volume of oxygen is 32 litres?

15.41 In the electrolysis of a chloride ion solution, a current of 0.15 A was passed for 30 minutes. What volume of chlorine gas would be given off at the positive electrode if the molar volume of chlorine is 21 litres? (Assume that chlorine is the only gas evolved at this electrode.)

15.42 In the electrolysis of an aqueous solution, a current of 0.5 A was passed for 50 minutes. What volume of oxygen gas would be given off at the positive electrode if the molar volume of oxygen is 20 litres?

15.43 In the electrolysis of an aqueous solution, a current of 0.3 A was passed for 40 minutes. What volume of hydrogen gas would be given off at the negative electrode if the molar volume of hydrogen is 25 litres?

15.44 In the electrolysis of an alkali solution, a current of 0.25 A was passed for 2 hours. What volume of oxygen gas would be given off at the positive electrode if the molar volume of oxygen is 19 litres?

15.45 In the electrolysis of an aqueous solution, a current of 0.1 A was passed for 40 minutes. What volume of hydrogen gas would be given off at the negative electrode if the molar volume of hydrogen is 22 litres?

PART C: More difficult Problems

Worked Example 15.5 involves calculation of the time an electrolysis must have been taking place, given the amount of an element which is produced at an electrode. Problems 15.46–15.60 are of this type, either requiring the time or the current flowing to be calculated.

Worked Example 15.5

For how long, in minutes, must a current of 0.25 A flow in the electrolysis of molten aluminium oxide to cause the deposition of 1.08 g of aluminium at the negative electrode?

In this type of example, instead of being told current and time, we have to calculate one of these quantities, having been given information about the amount of product at an electrode. We start by calculating how many moles of aluminium have been deposited.

$$27 \ \text{g} = 1 \ \text{mol of Al}$$

$$1 \ \text{g} = \frac{1}{27} \ \text{mol of Al}$$

$$1.08 \ \text{g} = \frac{1.08}{27} \ \text{mol of Al}$$

$$= 0.04 \ \text{mol of Al}$$

We next have to find out how many moles of electrons must have been involved. The ion-electron equation for the electrode reaction must be written:

$$Al^{3+} + 3e^- \rightarrow Al$$

This equation tells us that:

$$1 \ \text{mol of Al is deposited by 3 mol of electrons}$$

$$0.04 \ \text{mol of Al is deposited by } 0.04 \times 3 \ \text{mol of electrons}$$

$$= \textbf{0.12 mol of electrons}$$

We now find what quantity of charge this represents:

$$
\begin{aligned}
1 \quad \text{mol of electrons} &= \text{a charge of } 96\,500 \quad \text{C} \\
0.12 \quad \text{mol of electrons} &= \text{a charge of } 0.12 \times 96\,500 \text{ C} \\
&= 11\,580\,\text{C}
\end{aligned}
$$

Since we know that the total charge, current and time are connected by the equation $Q = I \times t$, we can now use the total charge (calculated above) and the known current to calculate the time.

$$
\begin{aligned}
Q &= I \times t \\
11\,580 &= 0.25 \times t
\end{aligned}
$$

So

$$
\begin{aligned}
t &= \frac{11\,580}{0.25}\,\text{s} \\
&= 46\,320\,\text{s} \\
&= 772 \text{ minutes}
\end{aligned}
$$

PROBLEMS

Problems 15.46–15.60 are of the type illustrated by Worked Example 15.5, involving the calculation of time or current flowing in an electrolysis.

15.46 6.335 g of copper was deposited at the negative electrode in the electrolysis of copper(II) sulphate solution using a current of 0.5 A. For how long, in minutes, had the current been flowing?

15.47 An aqueous solution was electrolysed for 20 minutes during which 30 cm³ of oxygen was evolved at the positive electrode according to the equation:

$$2H_2O\,(l) \longrightarrow O_2\,(g) + 4H^+\,(aq) + 4e^-$$

If the molar volume of oxygen under the experimental conditions was 24 litres, what current must have been flowing?

15.48 In the electrolysis of silver(I) nitrate solution, 1.079 g of silver was deposited in 1 hour. What current must have been flowing?

15.49 When potassium hydroxide solution is electrolysed, the equation for the reaction at the negative electrode is:

$$2H_2O\,(l) + 2e^- \longrightarrow H_2\,(g) + 2OH^-\,(aq)$$

A solution of potassium hydroxide was electrolysed using a current of 0.1 A. If 0.06 g of hydrogen was evolved, for how long, in minutes, had the current been flowing?

15.50 0.3175 g of copper was deposited in the electrolysis of a solution of copper(II) nitrate. For how long, in minutes, must the current of 0.5 A have been flowing?

15.51 A solution of nickel(II) sulphate was electrolysed for 20 minutes, during which time 1.174 g of nickel was deposited. What current must have been flowing?

15.52 An aqueous solution was electrolysed for 30 minutes, during which time 165 cm³ of oxygen was evolved at the positive electrode according to the equation:

$$2H_2O \,(l) \longrightarrow O_2 \,(g) \,+\, 4H^+ \,(aq) \,+\, 4e^-$$

If the molar volume of oxygen under experimental conditions was 22 litres, what current must have been flowing?

15.53 Nickel(II) nitrate solution is electrolysed using a current of 0.2 A, yielding 1.174 g of nickel. For how long, in minutes, had the current been flowing?

15.54 A solution of chromium(III) sulphate is electrolysed for 1 hour, during which 0.624 g of chromium is deposited. What current must have been flowing?

15.55 When dilute sulphuric acid is electrolysed, the evolution of oxygen at the positive electrode is described by the following ion-electron equation:

$$2H_2O \,(l) \longrightarrow O_2 \,(g) \,+\, 4H^+ \,(aq) \,+\, 4e^-$$

Dilute sulphuric acid is electrolysed using a current of 0.12 A. If 0.48 g of oxygen is evolved, for how long, in minutes, must the current have been flowing?

15.56 In the industrial electrolysis of a nickel(II) solution to plate metal objects with a layer of nickel, a current of 500 A was used to deposit 11.74 kg of nickel. How long, in hours, had the electrolysis been running?

15.57 4.16 kg of chromium was plated during three hours of electrolysis of a chromium(III) compound. What current must have been used?

15.58 For how long, in minutes, must a current of 1.2×10^5 A be applied to obtain 100 kg of aluminium in the industrial electrolysis of aluminium oxide?

15.59 154 kg of sodium was obtained in 10 hours of electrolysis of a molten sodium salt. What current must have been flowing during this time?

15.60 12.7 kg of copper was deposited during four hours of electrolytic purification of copper using a solution of copper(II) sulphate. What current must have been applied?

Worked Example 15.6

A molten iron compound is electrolysed using a current of 4.73 A for 30 minutes during which 1.64 g of iron is deposited. Calculate the number of positive charges on each iron ion.

In this type of problem we do not know whether the compound is an iron(II) or iron(III) compound which is being electrolysed; that is, we do not know whether it is Fe^{2+} or Fe^{3+} ions which are being discharged. We therefore cannot write the ion-electron equation as we have before. We need to work out the ion-electron equation by calculating the number of moles of electrons which have been passed and then the number of moles of iron which have been deposited.

$$Q = It$$
$$= 4.73 \times 30 \times 60 \ C$$
$$= 8514 \ C$$

$$96\,500 \ C = 1 \quad \text{mol of electron}$$
$$1 \quad C = \frac{1}{96\,500} \ \text{mol of electrons}$$
$$8514 \quad C = \frac{8514}{96\,500} \ \text{mol of electrons}$$
$$= 0.0882 \ \text{mol of electrons (rounded)}$$

$$55.8 \quad g = 1 \quad \text{mol of iron}$$
$$1 \quad g = \frac{1}{55.8} \quad \text{mol of iron}$$
$$1.64 \quad g = \frac{1.64}{55.8} \quad \text{mol of iron}$$
$$= 0.0294 \ \text{mol of iron (rounded)}$$

What we have now calculated is that:

0.0294 mol of Fe was deposited by 0.0882 mol of electrons

So

$$1 \text{ mol of Fe would be deposited by } \frac{0.0882}{0.0294} \text{ mol of electrons}$$

$$= 3 \text{ mol of electrons}$$

Thus the charge on the iron ion must be 3+ and the ion-electron equation is:

$$Fe^{3+} + 3e^- \longrightarrow Fe$$

Note that the figures will not always turn out to be exact whole number multiples since the data are usually obtained by experiment and there will be some error as a result. Also, any 'rounding' taking place during calculation may contribute to a difference between the calculated relationship and the exact whole-number ratio that we are looking for.

PROBLEMS

Problems 15.61–15.65 are of the type illustrated in Worked Example 15.6.

15.61 A molten metal oxide is electrolysed using a current of 0.2 A for one hour, during which 0.322 g of the metal, relative atomic mass 87.6, is deposited at the negative electrode. Calculate the number of charges on each metal ion.

15.62 A chromium compound was electrolysed using a current of 0.2 A for 45 minutes, during which 0.097 g of chromium was deposited. What was the charge on a chromium ion in this compound?

15.63 A molten vanadium compound is electrolysed using a current of 0.15 A for 40 minutes, during which 0.0634 g of vanadium was deposited. Calculate the charge on the vanadium ions in this compound.

15.64 0.304 g of a metal X, relative atomic mass 39.1, was obtained from the electrolysis of molten X chloride after a current of 0.25 A had been flowing for 50 minutes. Calculate the charge on the ions of metal X.

15.65 A tin compound is electrolysed using a current of 0.4 A for one hour, during which 0.886 g of tin is deposited. Calculate the charge on each tin ion in the compound.

Radioactive Half-life

When a radioactive isotope emits radiation, the intensity of the radiation decreases with time, according to the graph shown below.

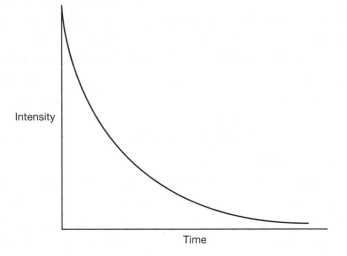

A feature of radioactive decay is that a quantity known as **half-life** (symbol $t_{1/2}$), which is a constant for any specified radioisotope, can be measured. This is the time during which the intensity of the radiation decreases to **half** what it was at the start of timing. Half-life values vary from fractions of a second to many millions of years; for example the half-life of ^{213}Po is 4.2×10^{-6} seconds, while that of ^{40}K is 1.3×10^{9} years.

A simple illustration of how half-life can be calculated is shown in the graph below.

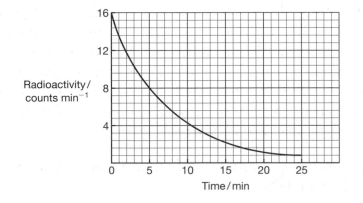

As can be seen, at the start of timing (time = 0), the radioactivity has a value of 16 counts min^{-1}. After 5 minutes, the count rate has dropped to 8 counts min^{-1}, i.e. to **half** the original value. After a further 5 minutes, the count rate has dropped from 8 to 4 counts min^{-1}, again a halving of the radioactivity. It can be seen from the graph that in ANY period of 5 minutes, the radioactivity decreases by half. The value of 5 minutes is thus the **half-life** for this particular radioisotope.

Half-life problems involve three quantities:

(a) **The half-life.** As mentioned before, this is a constant for the particular radioisotope and is entirely independent of how old the sample is, or whether the radioisotope is in the form of the pure element or in a compound.

(b) **The time over which the decay of the radioisotope has been measured.** This can be expressed in the usual units of time, e.g. 20 minutes, 500 years, etc. It can also be expressed as a number of half-lives, since the half-life for the radioisotope is an actual amount of time.

(c) **The quantity of the radioisotope or the intensity of the radiation.** This can be expressed in a number of ways:

- If the quantity of the radioisotope (or the intensity of the radiation) at the start of timing is referred to as 1 or 100% (i.e. the original or whole amount), we can refer to the sample having decayed to a fraction or a percentage of its original value, e.g. $\frac{1}{2}$ (50%), $\frac{1}{4}$ (25%) etc.

- The intensity of the radiation, as measured by a Geiger counter, is given in counts per minute (counts min^{-1}). This is proportional to the quantity of the radioactive isotope present.

- The mass of the radioisotope present. Care is needed in using mass, since, although the **mass of the original radioisotope** is decreasing, the **total mass of the sample** may not change much or at all. For example, when radioactive ^{209}Pb atoms decay, they do so by β-emission and become atoms of ^{209}Bi. If we started with a 1 g sample of ^{209}Pb, we would eventually end up with a 1 g sample of ^{209}Bi, since β-emission involves (virtually) no loss of mass. The mass of radioactive Pb would have been constantly decreasing, but the total mass of the sample would have remained the same. With α-emission, a small loss of total mass takes place. For example, ^{210}Po is an α-emitter which turns into ^{206}Pb. A 210 g sample of ^{210}Po would decay eventually into 206 g of ^{206}Pb, the other 4 g having been lost as α-particles.

How to approach problems involving half-life can best be seen by considering the following Worked Examples.

Worked Example 16.1

A radioisotope has a half-life of 5 minutes. How long will it take for the radioactive count rate to drop to 12.5% of its original value?

Since we are given a **percentage** value of the original isotope, we describe the original quantity as being 100% (i.e. the whole amount). We proceed as follows.

At start	we have 100%	of the original count rate
After 1 half-life	we have 50%	of the original count rate
After 2 half-lives	we have 25%	of the original count rate
After 3 half-lives	we have 12.5%	of the original count rate.

So, in a period of time of 3 half-lives, the count rate has dropped to 12.5% of its original value. Since we are told that the half-life is 5 minutes, we conclude that the sample has been decaying for $3 \times 5 = \mathbf{15\ minutes}$.

Worked Example 16.2

A radioisotope has a half-life of 3 days. What fraction of the original isotope will be present if a pure sample of it decays for 12 days?

The sample has a half-life of 3 days and has been allowed to decay for 12 days; this is a time of **4 half-lives**.

Since we are asked for a **fraction** of the original isotope, we describe the original quantity as being 1 (i.e. the whole amount). We proceed as follows.

At start	we have 1 (the original amount)
After 1 half-life	we have $\frac{1}{2}$ of the original amount
After 2 half-lives	we have $\frac{1}{4}$ of the original amount
After 3 half-lives	we have $\frac{1}{8}$ of the original amount
After 4 half-lives	we have $\frac{1}{16}$ of the original amount

So there will be $\frac{1}{16}$th of the original radioisotope left after 4 half-lives.

Worked Example 16.3

A sample of a radioisotope has a count rate, measured by a Geiger counter, of 48 counts per minute. 6 hours later, the count rate is 3 counts min^{-1} (counts per minute). What is the half-life of the radioisotope?

Since we are given the initial 'amount' of the radioisotope in terms of its count rate, we proceed as follows.

At start	the count rate is 48 counts min^{-1}
After 1 half-life	the count rate is 24 counts min^{-1}
After 2 half-lives	the count rate is 12 counts min^{-1}
After 3 half-lives	the count rate is 6 counts min^{-1}
After 4 half-lives	the count rate is 3 counts min^{-1}

In a period of time of 4 half-lives, the count rate has dropped to 3 counts per minute. But we are told that this period of time is 6 hours.

$$\text{So } \mathbf{1} \text{ half-life must be } \frac{6}{4} = \mathbf{1.5 \text{ hours.}}$$

PROBLEMS

The following problems are of the types shown in Worked Examples 16.1 to 16.3.

16.1 A sample of a radioisotope decays to $\frac{1}{8}$th of its original activity over a period of 12 days. What is the half-life of the radioisotope?

16.2 A radioisotope has a half-life of 14 hours. How long will it take to decay to 6.25% of its original activity?

16.3 A radioisotope with a half-life of 9 seconds has a count rate of 64 counts min^{-1} at the start of timing. What will the count rate be 18 seconds later?

16.4 A sample of a radioisotope decays to $\frac{1}{32}$ of its original activity in 12.5 years. What is the half-life of the isotope?

16.5 Over what period of time would a radioisotope with a half-life of 8.3 days decay to 12.5% of its original activity?

16.6 What fraction of its original activity will a sample of a radioisotope with a half-life of 3 minutes have after decaying for 15 minutes?

16.7 A radioisotope decays to $\frac{1}{16}$ of its original activity over a period of 20 minutes. What is its half-life?

16.8 Over what period of time would a radioisotope, with a half-life of 7 seconds, decay in activity from 80 to 10 counts min^{-1}?

16.9 A radioisotope has a half-life of 160 days. A 1 g sample of the pure radioisotope is allowed to decay for 640 days. What mass of the original isotope will remain after that time?

16.10 A radioisotope decays from an activity of 56 to 7 counts per minute over a period of 3.6 hours. What is the half-life of the radioisotope?

16.11 ^{257}Lr has a half-life of 8 seconds. Over what period of time would the activity of a sample of this radioisotope decay from 120 to 15 counts min^{-1}?

16.12 ^{253}No has a half-life of 10 minutes. What percentage of this radioisotope would remain after a sample had decayed for a period of 20 minutes?

16.13 A sample of the radioisotope ^{24}Na decays to $\frac{1}{32}$ of its original activity over a period of 75 hours. What is the half-life of ^{24}Na?

16.14 A sample of ^{254}Es, with a half-life of 280 days, has a count rate of 96 counts min^{-1}. How long will it take for the activity to decay to 12 counts min^{-1}?

16.15 ^{90}Sr has a half-life of 27 years. What fraction of a sample of this radioisotope would remain after 108 years?

16.16 Tritium, ^{3}H, one of the heavy isotopes of hydrogen, decays to $\frac{1}{64}$ of its original activity in 73.8 years. What is the half-life of this isotope?

16.17 ^{210}At has a half-life of 8.3 hours. Over what period of time would its activity decay to $\frac{1}{64}$ of its original value?

16.18 ^{253}Fm has a half-life of 4.5 days. What percentage of a sample of this radioisotope would remain after decaying for 22.5 days?

16.19 ^{36}Cl decays to 6.25% of its original value in 1.2×10^{6} years. What is the half-life of this isotope?

16.20 ^{40}K has a half-life of 1.3×10^{9} years. After what period time would its activity decay to 6.25% of its original value?

Additional Theory: Dating Old Materials Using Carbon-14

A certain amount of radioactive carbon-14 is formed continuously in the upper atmosphere and is taken in by plants in the form of carbon dioxide during photosynthesis. Animals which eat these plants also absorb this radioactivity. It is thought that the level of radioactivity in the upper atmosphere has been constant for thousands of years, so we can assume that the level of radioactivity in a plant growing thousands of years ago is the same as that of a plant growing today. When a plant dies, or is eaten by an animal, it no longer absorbs carbon dioxide from the air and the amount of radioactivity in it decays. Since we know the half-life of carbon-14 (5570 years), we can use this fact to determine the age of animal and plant remains many thousands of years old. This is known as carbon dating. This is best illustrated by the following Worked Example.

Worked Example 16.4

An ancient animal skin coat is analysed for radioactivity from carbon-14; this level is found to be 12.5% of the level of radioactivity from a living plant. How old is the ancient animal skin coat?

We assume that the level of radioactivity from the animal skin coat, **at the time the animal was alive,** was the same as the level in animals and plants alive today. Since the level of radioactivity is expressed as a percentage of the original value, we proceed as follows.

When the animal was alive the count rate would be 100% of that of living things today.

After 1 half-life   the count rate would be 50% of that of living things today.

After 2 half-lives   the count rate would be 25% of that of living things today.

After 3 half-lives   the count rate would be 12.5% of that of living things today.

In a period of time of **3 half-lives,** the level of radioactivity in the animal skin coat would have decayed to 12.5% of that of current animal or plant material. But we know that the half-life of carbon-14 is 5570 years so we can say that the age of the coat is:

$$3 \times 5570 = 16\,710 \text{ years}$$

PROBLEMS

The following problems involve the use of 'carbon dating' illustrated in Worked Example 16.4.

In each of these problems take the half-life of carbon-14 to be 5570 years.

16.21 A sample of ancient wood is found to have a radioactive count rate due to carbon-14 of 12 counts min^{-1}. A sample of modern wood has a count rate of 48 counts min^{-1}. Calculate the age of the sample of ancient wood.

16.22 A leather garment is calculated to be 16 710 years old, using carbon dating. If a similar sample of modern leather has a carbon-14 count rate of 32 counts min^{-1}, what will the count rate from the ancient leather sample be?

16.23 A sample of ancient charred wood found in a Stone Age cave has a count rate of $\frac{1}{32}$ of that of modern wood. Calculate the age of the charred wood.

16.24 A sample of bone found in a prehistoric dwelling has a carbon-14 count rate which is 12.5% of that of a similar sample of bone in a living person. Calculate the age of the prehistoric bone.

16.25 A fossilised tree sample has a count rate of 6 counts min^{-1} due to carbon-14. A similar sample cut from a living tree has a count rate of 96 counts min^{-1}. Calculate the age of the fossil.

Appendix 1

Table of Relative Atomic Masses

Element	Symbol	RAM	Element	Symbol	RAM
Aluminium	Al	27.0	Magnesium	Mg	24.3
Argon	Ar	40.0	Mercury	Hg	200.6
Barium	Ba	137.3	Neon	Ne	20.2
Bromine	Br	79.9	Nickel	Ni	58.7
Calcium	Ca	40.0	Nitrogen	N	14.0
Carbon	C	12.0	Oxygen	O	16.0
Chlorine	Cl	35.5	Phosphorus	P	31.0
Chromium	Cr	52.0	Platinum	Pt	195.1
Cobalt	Co	58.9	Potassium	K	39.1
Copper	Cu	63.5	Silicon	Si	28.1
Fluorine	F	19.0	Silver	Ag	107.9
Gold	Au	197.0	Sodium	Na	23.0
Helium	He	4.0	Sulphur	S	32.1
Hydrogen	H	1.0	Tin	Sn	118.7
Iodine	I	126.9	Titanium	Ti	47.9
Iron	Fe	55.8	Vanadium	V	51.0
Krypton	Kr	83.8	Xenon	Xe	131.3
Lead	Pb	207.2	Zinc	Zn	65.4
Lithium	Li	6.9			

Appendix 2

Enthalpies of Formation and Combustion of Selected Substances

Substance	Standard enthalpy of formation $kJ\ mol^{-1}$	Standard enthalpy of combustion $kJ\ mol^{-1}$
hydrogen	–	−286
carbon (graphite)	–	−394
sulphur (rhombic)	–	−297
methane	−75	−891
ethane	−85	−1560
propane	−104	−2220
butane	−125	−2877
benzene	49	−3268
ethene	52	−1411
ethyne	227	−1300
methanol	−239	−727
ethanol	−278	−1367
propan-1-ol	−306	−2020
methanoic acid	−409	−255
ethanoic acid	−487	−876

Electrolysis Of Water

Reduction reactions at the negative electrode

$2H_2O\ (l) + 2e^- \longrightarrow H_2\ (g) + 2OH^-\ (aq)$

$2H^+\ (aq) + 2e^- \longrightarrow H_2\ (g)$

Oxidation reactions at the positive electrode

$2H_2O\ (l) \longrightarrow O_2\ (g) + 4H^+\ (aq) + 4e^-$

$4OH^-\ (aq) \longrightarrow 2H_2O\ (l) + O_2\ (g) + 4e^-$

Answers

CHAPTER 1 – Answers

1.1	32.8 g	**1.2**	0.02 mol	**1.3**	1.2 mol
1.4	8.55 g	**1.5**	0.0025 mol	**1.6**	59.6 g
1.7	0.3 mol	**1.8**	2.24 g	**1.9**	0.02 mol
1.10	48 g	**1.11**	26.9 g	**1.12**	1.5 mol
1.13	3.99 g	**1.14**	0.04 mol	**1.15**	147 g
1.16	0.8 mol	**1.17**	124.8 g	**1.18**	0.05 mol
1.19	31.8 g	**1.20**	0.03 mol	**1.21**	$0.2 \text{ mol } l^{-1}$
1.22	0.3 mol	**1.23**	$0.2 \text{ mol } l^{-1}$	**1.24**	40 cm^3
1.25	0.15 mol	**1.26**	$0.2 \text{ mol } l^{-1}$	**1.27**	250 cm^3
1.28	32 g	**1.29**	$0.25 \text{ mol } l^{-1}$	**1.30**	1.321 g
1.31	1.5 l	**1.32**	40 cm^3	**1.33**	0.5 l
1.34	7.98 g	**1.35**	$0.05 \text{ mol } l^{-1}$	**1.36**	$0.05 \text{ mol } l^{-1}$
1.37	$0.1 \text{ mol } l^{-1}$	**1.38**	0.96 g	**1.39**	$0.2 \text{ mol } l^{-1}$
1.40	25 cm^3				

CHAPTER 2 – Answers

2.1	0.5 g	**2.2**	12.7 g
2.3	0.6 g	**2.4**	3.5 g
2.5	11.5 g	**2.6**	0.4 g
2.7	7.29 g	**2.8**	2.2 g
2.9	66 g	**2.10**	1.6 g
2.11	1282 kg	**2.12**	37.5 kg
2.13	44.64 tonne	**2.14**	9612 kg
2.15	672 kg	**2.16**	2.76×10^3 kg
2.17	3.52×10^4 kg	**2.18**	3.68 tonne
2.19	1.9×10^5 kg	**2.20**	8.8×10^3 kg

CHAPTER 3 – Answers

3.1	$0.5 \text{ mol } l^{-1}$	**3.2**	16 cm^3
3.3	6.91 g	**3.4**	300 cm^3
3.5	$0.8 \text{ mol } l^{-1}$	**3.6**	4.15 g
3.7	$0.182 \text{ mol } l^{-1}$	**3.8**	1.167 g
3.9	$0.094 \text{ mol } l^{-1}$	**3.10**	$0.2 \text{ mol } l^{-1}$
3.11	$0.0584 \text{ mol } l^{-1}$	**3.12**	$0.2 \text{ mol } l^{-1}$
3.13	$0.36 \text{ mol } l^{-1}$	**3.14**	$0.12 \text{ mol } l^{-1}$
3.15	$0.021 \text{ mol } l^{-1}$	**3.16**	*(a)* $0.2 \text{ mol } l^{-1}$; *(b)* 10 times; *(c)* $2 \text{ mol } l^{-1}$
3.17	*(a)* $2 \text{ mol } l^{-1}$; *(b)* 100 cm^3	**3.18**	*(a)* $0.2 \text{ mol } l^{-1}$; *(b)* $0.32 \text{ mol } l^{-1}$
3.19	*(a)* 100 times; *(b)* $0.025 \text{ mol } l^{-1}$; *(c)* $2.5 \text{ mol } l^{-1}$		
3.20	*(a)* 0.02 mol; *(b)* $0.04 \text{ mol } l^{-1}$; *(c)* $0.04 \text{ mol } l^{-1}$		

CHAPTER 4 – Answers

4.1 *(a)* $0.0005\,g\,s^{-1}$ $(5 \times 10^{-4}\,g\,s^{-1})$ *(b)* $0.00015\,g\,s^{-1}$ $(1.5 \times 10^{-4}\,g\,s^{-1})$

4.2 *(a)* $0.00375\,mol\,l^{-1}\,s^{-1}$ $(3.75 \times 10^{-3}\,mol\,l^{-1}\,s^{-1})$ *(b)* $0.00025\,mol\,l^{-1}\,s^{-1}$ $(2.5 \times 10^{-4}\,mol\,l^{-1}\,s^{-1})$

4.3 *(a)* $1.1\,cm^3\,s^{-1}$ *(b)* $0.2\,cm^3\,s^{-1}$

4.4 *(a)* $0.002\,mol\,l^{-1}\,s^{-1}$ $(2 \times 10^{-3}\,mol\,l^{-1}\,s^{-1})$ *(b)* $0.0006\,mol\,l^{-1}\,s^{-1}$ $(6 \times 10^{-4}\,mol\,l^{-1}\,s^{-1})$

4.5 *(a)* $0.0125\,g\,s^{-1}$ *(b)* $0.0025\,g\,s^{-1}$ $(2.5 \times 10^{-3}\,g\,s^{-1})$

4.6 *(a)* $0.005\,cm^3\,s^{-1}$ $(5 \times 10^{-4}\,cm^3\,s^{-1})$ *(b)* $0.000064\,cm^3\,s^{-1}$ $(6.4 \times 10^{-5}\,cm^3\,s^{-1})$

4.7 *(a)* $0.002\,mol\,l^{-1}\,min^{-1}$ $(2 \times 10^{-3}\,mol\,l^{-1}\,min^{-1})$ *(b)* $0.0004\,mol\,l^{-1}\,min^{-1}$ $(4 \times 10^{-4}\,mol\,l^{-1}\,min^{-1})$

4.8 *(a)* $0.0025\,g\,s^{-1}$ $(2.5 \times 10^{-3}\,g\,s^{-1})$ *(b)* $0.003\,g\,s^{-1}$ $(3 \times 10^{-3}\,g\,s^{-1})$

4.9 *(a)* $0.5\,cm^3\,s^{-1}$ *(b)* $0.1\,cm^3\,s^{-1}$

4.10 *(a)* $0.000096\,mol\,l^{-1}\,s^{-1}$ $(9.6 \times 10^{-5}\,mol\,l^{-1}\,s^{-1})$ *(b)* $0.00005\,mol\,l^{-1}\,s^{-1}$ $(5 \times 10^{-5}\,mol\,l^{-1}\,s^{-1})$

4.11 *(a)* $0.37\,mol\,l^{-1}$ *(b)* $50\,s$

4.12 *(a)* $25\,s$ *(b)* $33\,°C$

     (c) $0.004\,s^{-1}$, $0.008\,s^{-1}$, $0.016\,s^{-1}$ and $0.032\,s^{-1}$ respectively. The rate doubles with each $10\,°C$ rise in temperature.

4.13 *(a)* $0.32\,mol\,l^{-1}$ *(b)* $40\,s$

4.14 *(a)* $25\,°C$ *(b)* $25\,s$

4.15 *(a)* $62.5\,s$ *(b)* $0.0084\,mol\,l^{-1}$

CHAPTER 5 – Answers

Answers have, in some cases, been rounded to 3 significant figures.

5.1 *(a)* Pb in excess; *(b)* $0.05\,g$ **5.2** *(a)* $CaCO_3$ in excess; *(b)* $4.4\,g$

5.3 *(a)* KI in excess; *(b)* $11.1\,g$ **5.4** *(a)* Na_2CO_3 in excess; *(b)* $1.1\,g$

5.5 *(a)* NaOH in excess; *(b)* $1.56\,g$ **5.6** *(a)* $MgCl_2$ in excess; *(b)* $0.258\,g$

5.7 *(a)* $BaCl_2$ in excess; *(b)* $0.934\,g$ **5.8** *(a)* $(NH_4)_2SO_4$ in excess; *(b)* $1.02\,g$

5.9 *(a)* FeS in excess; *(b)* $1.36\,g$ **5.10** *(a)* CuO in excess; *(b)* $0.636\,g$; *(c)* $1.92\,g$

CHAPTER 6 – Answers

Answers have been rounded to a **maximum** of 3 significant figures.

Note: In Problems 6.1 to 6.5, the questions state whether heat is given out or absorbed, and simply ask for the amount of heat. So, strictly speaking, the answers do not need a + or − sign, as the direction of the enthalpy change is given in the question. However, signs are given in these answers, to emphasise whether the process is exothermic or endothermic.

6.1	$-3.34\,kJ$	**6.2**	$-3.34\,kJ$	**6.3**	$+2.09\,kJ$
6.4	$-1.25\,kJ$	**6.5**	$-39.7\,kJ$	**6.6**	$+167\,kJ\,mol^{-1}$
6.7	$-188\,kJ\,mol^{-1}$	**6.8**	$-334\,kJ\,mol^{-1}$	**6.9**	$-167\,kJ\,mol^{-1}$
6.10	$+62.7\,kJ\,mol^{-1}$	**6.11**	$-669\,kJ\,mol^{-1}$	**6.12**	$-25.1\,kJ\,mol^{-1}$
6.13	$-836\,kJ\,mol^{-1}$	**6.14**	$+25.1\,kJ\,mol^{-1}$	**6.15**	$-1340\,kJ\,mol^{-1}$
6.16	$-41.8\,kJ\,mol^{-1}$	**6.17**	$-2090\,kJ\,mol^{-1}$	**6.18**	$+16.7\,kJ\,mol^{-1}$
6.19	$-2660\,kJ\,mol^{-1}$	**6.20**	$+33.4\,kJ\,mol^{-1}$	**6.21**	$3.92\,g$
6.22	$0.422\,g$	**6.23**	$0.295\,°C$	**6.24**	$0.335\,g$
6.25	$6.6\,°C$ rise	**6.26**	$10.7\,°C$	**6.27**	$5.96\,g$
6.28	$33.1\,g$	**6.29**	$11.1\,g$	**6.30**	$13.5\,g$
6.31	$-58.5\,kJ\,mol^{-1}$	**6.32**	$-57.7\,kJ\,mol^{-1}$	**6.33**	$-56.8\,kJ\,mol^{-1}$
6.34	$-56.8\,kJ\,mol^{-1}$	**6.35**	$-57.1\,kJ\,mol^{-1}$	**6.36**	$3.43\,°C$
6.37	$6.85\,°C$	**6.38**	$3.43\,°C$	**6.39**	$3.43\,°C$
6.40	$9.14\,°C$				

CHAPTER 7 – Answers

The answers below are expressed to a **maximum** of 3 significant figures.

7.1	1.20×10^{24}	**7.2**	1.81×10^{24}
7.3	7.22×10^{23}	**7.4**	0.002
7.5	0.0375	**7.6**	1.20×10^{23}
7.7	1.2×10^{23}	**7.8**	0.6
7.9	9.03×10^{23}	**7.10**	0.04
7.11	2.41×10^{23}	**7.12**	1×10^{-4}
7.13	4.06×10^{24}	**7.14**	2
7.15	7.22×10^{23}	**7.16**	3.3
7.17	4.33×10^{24}	**7.18**	2.71×10^{23}
7.19	0.05 mol	**7.20**	1.44×10^{23}
7.21	6.02×10^{21}	**7.22**	0.3 g
7.23	3.01×10^{22}	**7.24**	0.416 g
7.25	3.27×10^{-22} g	**7.26**	1.20×10^{23}
7.27	0.0486 g	**7.28**	2.41×10^{24}
7.29	0.14 g	**7.30**	7.31×10^{-23} g
7.31	1.08×10^{23}	**7.32**	0.0025 g
7.33	7.22×10^{22}	**7.34**	0.32 g
7.35	(a) 9.63×10^{22}; (b) 4.82×10^{22}	**7.36**	7.97×10^{-23} g
7.37	9.63×10^{22}	**7.38**	9.03×10^{22}
7.39	7.22×10^{23}	**7.40**	0.068 g

CHAPTER 8 – Answers

Note: The answers below are given to a **maximum** of three significant figures.

8.1	$22.2\,l$	**8.2**	$1.28\,g\,l^{-1}$
8.3	N_2: $0.237\,l$; H_2: $0.234\,l$	**8.4**	15.9
8.5	$0.525\,g\,l^{-1}$	**8.6**	$22.4\,l$
8.7	64.0	**8.8**	$0.52\,g\,l^{-1}$
8.9	30.0; ethane, C_2H_6	**8.10**	32.1; oxygen, O_2
8.11	$22.4\,l$	**8.12**	0.0236 g
8.13	$3.24\,l$	**8.14**	$100\,l$
8.15	185 g	**8.16**	$0.0449\,l$
8.17	$24.3\,l$	**8.18**	1100 g
8.19	$0.0248\,l$	**8.20**	$24.3\,l$
8.21	(a) $41.2\,l$; (b) O_2: $0.777\,g\,l^{-1}$; CO_2: $1.07\,g\,l^{-1}$	**8.22**	$24\,l$; 38.0
8.23	15.8	**8.24**	31.8
8.25	0.0802 g		

CHAPTER 9 – Answers

9.1	(a) O_2; (b) $100\,cm^2\,O_2$ and $100\,cm^3\,CO_2$	**9.2**	(a) CO; (b) $40\,cm^3\,CO_2$ and $10\,cm^3\,CO$
9.3	$60\,l\,CO_2$ and $40\,l\,O_2$	**9.4**	$1200\,l\,CO_2$ and $3200\,l\,O_2$
9.5	$400\,cm^3\,CO_2$, $600\,cm^3\,H_2O$ and $1300\,cm^3\,O_2$	**9.6**	$100\,l\,N_2$, $200\,l\,H_2O$ and $300\,l\,O_2$
9.7	$600\,l\,CO_2$, $1025\,l\,O_2$	**9.8**	(a) $2.4 \times 10^5\,l\,CO_2$; (b) $5.2 \times 10^5\,l\,H_2$
9.9	$5 \times 10^5\,l\,CCl_4$, $5 \times 10^5\,l\,S_2Cl_2$, $10^6\,l\,Cl_2$	**9.10**	$4.5 \times 10^5\,l$

CHAPTER 10 – Answers

10.1	90 l	**10.2**	1.1 l
10.3	272 g	**10.4**	320 l
10.5	10.9 g	**10.6**	4.4 l
10.7	40 000 l ($4 \times 10^4 l$)	**10.8**	798 g
10.9	3 l	**10.10**	234 kg

CHAPTER 11 – Answers

Note: The answers below have been rounded to a maximum of three significant figures.

11.1	11.2 g	**11.2**	13.4 g
11.3	0.45 g	**11.4**	33.5 g
11.5	5.75 g	**11.6**	7.14×10^4 kg
11.7	27.6 kg	**11.8**	7.20×10^4 kg
11.9	6.51×10^3 kg	**11.10**	666 kg
11.11	(a) 25.6 g; (b) 75%	**11.12**	40%
11.13	75%	**11.14**	75%
11.15	(a) 4.24 g; (b) 80%	**11.16**	60%
11.17	70%	**11.18**	60%
11.19	(a) 4.185 kg; (b) 90%	**11.20**	(a) 3.99×10^4 kg; (b) 60%

CHAPTER 12 – Answers

Note: In some cases, a calculated enthalpy change may differ slightly from the accepted value given in Appendix 2 or the SQA Data Book. This is because the original data are rounded to whole numbers and, in the calculation, any rounding error may be multiplied, resulting in a (very) small difference from the accepted value.

12.1	-891 kJ mol^{-1}	**12.2**	-1561 kJ mol^{-1}	**12.3**	-86 kJ mol^{-1}
12.4	-2222 kJ mol^{-1}	**12.5**	-129 kJ mol^{-1}	**12.6**	-1368 kJ mol^{-1}
12.7	-484 kJ mol^{-1}	**12.8**	-306 kJ mol^{-1}	**12.9**	-3271 kJ mol^{-1}
12.10	$+226$ kJ mol^{-1}	**12.11**	-425 kJ mol^{-1}	**12.12**	-491 kJ mol^{-1}
12.13	-137 kJ mol^{-1}	**12.14**	-79 kJ mol^{-1}	**12.15**	-312 kJ mol^{-1}
12.16	-412 kJ mol^{-1}	**12.17**	$+376$ kJ mol^{-1}	**12.18**	$+776$ kJ mol^{-1}
12.19	-409 kJ mol^{-1}	**12.20**	-5120 kJ mol^{-1}		

CHAPTER 13 – Answers

Note 1: Where concentrations are below 0.001 mol l^{-1} they are give only as a power of 10; concentrations of 0.001 mol l^{-1} or above are given both as a decimal fraction and as a power of 10. Where a concentration is 1 mol l^{-1}, this is given as 10^0 as well as 1, although in practice only the latter would be used.

Note 2: In Problems 13.31–13.40, where a dilution is by a factor of 100, the answers give the **two most obvious** ways of carrying out the dilution, in one or two steps. There will be **other** quite correct answers which are not given. The answers to such questions are marked with an asterisk (*).

13.1	3	**13.2**	6
13.3	5	**13.4**	0
13.5	4	**13.6**	8
13.7	11	**13.8**	4
13.9	7	**13.10**	14
13.11	10^{-5} mol l^{-1}	**13.12**	10^{-3} mol l^{-1} (or 0.001 mol l^{-1})

13.13 10^{-7} mol l^{-1}

13.14 10^{-13} mol l^{-1}

13.15 10^{-2} mol l^{-1} or 0.01 mol l^{-1}

13.16 10^{0} mol l^{-1} (more normally expressed as 1 mol l^{-1})

13.17 10^{-4} mol l^{-1}

13.18 10^{-10} mol l^{-1}

13.19 10^{-12} mol l^{-1}

13.20 10^{-1} mol l^{-1} or 0.1 mol l^{-1}

13.21 *(a)* 10^{-10} mol l^{-1}; *(b)* 10

13.22 *(a)* 10^{-6} mol l^{-1}; *(b)* 10^{-8} mol i^{-1}

13.23 *(a)* 10^{-12} mol l^{-1}; *(b)* 12

13.24 *(a)* 10^{-5} mol l^{-1}; *(b)* 10^{-9} mol l^{-1}

13.25 *(a)* 10^{-5} mol l^{-1}; *(b)* 5

13.26 *(a)* 10^{-14} mol l^{-1}; *(b)* 10^{0} mol l^{-1} (more normally expressed as 1 mol l^{-1})

13.27 *(a)* 10^{-11} mol l^{-1}; *(b)* 11

13.28 *(a)* 10^{-13} mol l^{-1}; *(b)* 10^{-1} mol l^{-1} (0.1 mol l^{-1})

13.29 *(a)* 10^{-1} mol l^{-1} (0.1 mol l^{-1}); *(b)* 1

13.30 *(a)* 10^{-3} mol l^{-1} (or 0.001 mol l^{-1}); *(b)* 10^{-11} mol l^{-1}

13.31 *(a)* 0; *(b)* take 10 cm^3 of original solution and dilute to 100 cm^3

13.32 *(a)* 2; *(b)* take 5 cm^3 of original solution and dilute to 50 cm^3

13.33 *(a)* 1; *(b)* take 25 cm^3 of original solution and dilute to 250 cm^3

13.34* *(a)* 1

(b) take 5 cm^3 of original solution and dilute to 500 cm^3 **or** take the whole 10 cm^3 of the original solution, dilute to 100 cm^3 and then take 50 cm^3 of this diluted solution and further dilute to 500 cm^3

13.35* *(a)* 0

(b) take 10 cm^3 of original solution and dilute to 1 litre **or** take 50 cm^3 of the original solution, dilute to 500 cm^3 and then take 100 cm^3 of this diluted solution and further dilute to 1 litre.

13.36 *(a)* 14; *(b)* take 5 cm^3 of original solution and dilute to 50 cm^3

13.37 *(a)* 13; *(b)* take 10 cm^3 of original solution and dilute to 100 cm^3

13.38 *(a)* 12; *(b)* take 25 cm^3 of original solution and dilute to 250 cm^3

13.39* *(a)* 14

(b) take 10 cm^3 of original solution and dilute to 1 litre **or** take 50 cm^3 of the original solution, dilute to 500 cm^3 and then take 100 cm^3 of this diluted solution and further dilute to 1 litre.

13.40* *(a)* 12

(b) take 5 cm^3 of original solution and dilute to 500 cm^3 **or** take the whole 10 cm^3 of the original solution, dilute to 100 cm^3 and then take 50 cm^3 of this diluted solution and further dilute to 500 cm^3.

CHAPTER 14 – Answers

14.1 50 cm^3

14.2 16 cm^3

14.3 0.0625 mol l^{-1}

14.4 40 cm^3

14.5 0.08 mol l^{-1}

14.6 0.05 mol l^{-1}

14.7 0.4 mol l^{-1}

14.8 18 cm^3

14.9 20 cm^3

14.10 38.4 cm^3

CHAPTER 15 – Answers

All answers have been rounded to three significant figures.

15.1 96,300 C

15.2 96,500 C

15.3 96,600 C

15.4 96,300 C

15.5 96,500 C

15.6 96,600 C

15.7 96,500 C

15.8 96,600 C

15.9 95,200 C

15.10 96,600 C

15.11 96,400 C

15.12 96,400 C

15.13 96,500 C

15.14 96,600 C

15.15 96,400 C

15.16 96,300 C

15.17 96,500 C

15.18 96,700 C

15.19	96,800 C	**15.20**	96,600 C	**15.21**	0.118 g
15.22	0.0554 g	**15.23**	0.302 g	**15.24**	0.00311 g
15.25	3.09 g				

15.26 *(a)* 0.00280 g H_2; *(b)* 0.0244 g O_2

15.27	0.168 g	**15.28**	*(a)* 0.00373 g H_2; *(b)* 0.0298 g O_2		
15.29	592 g	**15.30**	806 kg	**15.31**	42.9 kg
15.32	13.1 kg	**15.33**	725 kg	**15.34**	1.29 kg
15.35	376 kg	**15.36**	31.1 cm^3	**15.37**	239 cm^3
15.38	15.5 cm^3	**15.39**	23.3 cm^3	**15.40**	119 cm^3
15.41	29.4 cm^3	**15.42**	77.7 cm^3 (corrected)	**15.43**	93.3 cm^3
15.44	88.6 cm^3	**15.45**	27.4 cm^3	**15.46**	643 minutes
15.47	0.402 A	**15.48**	0.268 A	**15.49**	965 minutes (corrected)
15.50	32.2 minutes	**15.51**	3.22 A	**15.52**	1.61 A
15.53	322 minutes	**15.54**	0.965 A (corrected)	**15.55**	804 minutes
15.56	21.4 hours	**15.57**	2.14×10^3 A	**15.58**	149 minutes
15.59	1.79×10^4 A	**15.60**	2.68×10^3 A (corrected)	**15.61**	2+
15.62	3+	**15.63**	3+	**15.64**	1+
15.65	2+				

CHAPTER 16 – Answers

16.1	4 days	**16.2**	56 hours	**16.3**	16 counts min^{-1}
16.4	2.5 years	**16.5**	24.9 days	**16.6**	$\frac{1}{32}$
16.7	5 minutes	**16.8**	21 s	**16.9**	0.0625 g
16.10	1.2 hours	**16.11**	24 s	**16.12**	25%
16.13	15 hours	**16.14**	840 days	**16.15**	$\frac{1}{16}$
16.16	12.3 years	**16.17**	49.8 hours	**16.18**	3.125%
16.19	3×10^5 years	**16.20**	5.2×10^9 years	**16.21**	11,140 years
16.22	4 counts min^{-1}	**16.23**	27,850 years	**16.24**	16,710 years
16.25	22,280 years				